The Natural History of

DEER

Christopher Helm Mammal Series
Edited by Dr Ernest Neal, MBE, former President of the Mammal Society

Already published:
The Natural History of Antelopes
C.A. Spinage

The Natural History of Badgers
Ernest Neal

The Natural History of Otters
Paul Chanin

The Natural History of Squirrels
John Gurnell

The Natural History of Whales and Dolphins
P.G.H. Evans

Forthcoming titles include:

The Natural History of Moles
Martyn L. Gorman and David Stone

The Natural History of Rabbits
David Cowan

The Natural History of Seals
Nigel Bonner

The Natural History of Weasels and Stoats
Carolyn King

The Natural History of
DEER

RORY PUTMAN

Christopher Helm (Publishers) Ltd, Imperial House,
21–25 North Street, Bromley, Kent BR1 1SD

ISBN 0–7470–2603–3

A CIP catalogue record for this book is available from the British Library

Phototypeset by OPUS, Oxford
Printed and bound in Great Britain by Billing and Sons Ltd, Worcester

Contents

Colour plates

Figures

Tables

Acknowledgements

A book of this kind begins long before the author is ever aware of it. While my thanks must go to Ernest Neal for asking me to write the book itself, I owe an even bigger debt to those who first interested me in deer and who over the years have contributed piece by piece to whatever I may now claim to know of their biology. Gamekeepers, scientists, naturalists — all have taught me much, opened my eyes to facts, explanations and speculations I might never otherwise have seen. To all those alongside whom I have worked over the years, a very genuine thank you.

Within the context of the preparation of this particular book, however, I should also like to thank those who have contributed suggestions and ideas, those who have brought my attention to published material of which I was previously unaware, and those who have permitted me to quote previously unpublished data. I would mention particularly in such context Charles Santiapillai, Michael Green and Andrew Illius. Steven de Bie very generously allowed me to quote (pages 48–50) from an unpublished manuscript; Manuel Hinge kindly permitted reproduction here of Figure 3.1 from his PhD thesis, and provided the data on home ranges of red and roe deer quoted on pages 76–77; John Jackson permitted use of Figures 5.1 and 5.2 from his PhD thesis; Figure 3.3 and much of page 160 are based on information provided by Michael Green. Figure 5.3 and Tables 5.1 and 5.4 are taken from Dr H.R. Mishra's unpublished PhD thesis, while data presented on habitat use of chital, muntjac, sambar and hog deer in Chitawan National Park (pages 36–7) are also drawn from that thesis. Among my own students, I should like to thank Mags Cousins for the data included on page 164, Dawn Hawkins for the data on which is based Table 6.2, and Chris Mann: much of the material on the ecology of sika deer included in Chapter 3 is drawn from his unpublished PhD thesis.

Ernest Neal, Charles Santiapillai and my father each read the entire manuscript and made many useful comments for improvement. Various others read and commented on individual chapters or sections: George Darwall, Rory Harrington, Jochen Langbein, Simon Thirgood. To all of these my thanks.

For permission to reproduce here copyright material previously published elsewhere (figures, tables, sections of text), I thank Professor S.

Balasubramaniam, Dr C. Santiapillai, Dr M.R. Chambers and the editors of *Spixiana* (Table 3.5); Dr Tim Clutton-Brock, Dr Steve Albon and Miss Fiona Guinness (Figure 5.7 and textual quotes on page 83); Dr Chris Challies (Figure 8.3); George Darwall and Rob Clark (Figure 7.8 and textual quote pages 149–50); Dr Michael Green and the Zoological Society of London (Figure 3.8); Dr Tom Hanley (Figure 6.3); Dr Rory Harrington and the Royal Society of New Zealand (Table 3.4 and textual quotes on pages 32–3); Professor Reino Hofmann and the Royal Society of New Zealand (Figures 1.4 and 3.6); Dr Andrew Kitchener (Figure 7.4); Andrew Kitchener and the editors of *Modern Geology* (Figure 7.5); Dr Y.C. Kong and Dr P.H. But, together with the Royal Society of New Zealand, (Figures 8.1, 8.2 and text pages 158–9); and Dr Philip Ratcliffe (Figure 5.11). Finally, my thanks to Blackwell Scientific Publications for permission to reproduce Figure 2.1 from *A Field Guide to British Deer* (F.J. Taylor Page).

For preparation of original artwork, I thank Michael Clark, Tessa Lovatt-Smith and Raymond Cornick. A number of people very generously allowed me to use their photographs in the plates: I would like the thank Norma Chapman (Plate 3), Mags Cousins (Plate 22), Peter Gasson (jacket, Plates 7, 11, 12, 15 and 18), Jochen Langbein (Plates 21, 23 and 24), Chris Mann (Plate 5), and Bob Pratt (Plates 19 and 20); all other plates are by the author, with the exception of Plates 1 and 16 which appear courtesy of Bruce Coleman Ltd./Gerald Cubitt, Francisco Erize.

Finally, my sincere thanks to Dawn Trenchard for patiently converting my illegible scrawl into a finished typescript, to David Christie, my long-suffering copy-editor and to all at Christopher Helm who saw the book through to production.

Series editor's foreword

In recent years there has been a great upsurge of interest in wildlife and a deepening concern for nature conservation. For many there is a compelling urge to counterbalance some of the artificiality of present-day living with a more intimate involvement with the natural world. More people are coming to realise that we are all part of nature, not apart from it. There seems to be a greater desire to understand its complexities and appreciate its beauty.

This appreciation of wildlife and wild places has been greatly stimulated by the world-wide impact of natural-history television programmes. These have brought into our homes the sights and sounds both of our own countryside and of far-off places that arouse our interest and delight.

In parallel with this growth of interest there has been a great expansion of knowledge and, above all, understanding of the natural world — an understanding vital to any conservation measures that can be taken to safeguard it. More and more field workers have carried out painstaking studies of many species, analysing their intricate behaviour, relationships and the part they play in the general ecology of their habitats. To the time-honoured techniques of field observations and experimentation has been added the sophistication of radio-telemetry whereby individual animals can be followed, even in the dark and over long periods, and their activities recorded. Infra-red cameras and light-intensifying binoculars now add a new dimension to the study of nocturnal animals. Through such devices great advances have been made.

This series of volumes aims to bring this information together in an exciting and readable form so that all who are interested in wildlife may benefit from such a synthesis. Many of the titles in the series concern groups of related species such as otters, squirrels and rabbits so that readers from many parts of the world may learn about their own more familiar animals in a much wider context. Inevitably more emphasis will be given to particular species within a group as some have been more extensively studied than others. Authors too have their own special interests and experience and a text gains much in authority and vividness when there has been personal involvement.

Many natural history books have been published in recent years which have delighted the eye and fired the imagination. This is wholly good. But it is the intention of this series to take this a step further by exploring the subject in greater depth and by making available the results of recent research. In this way it is hoped to satisfy to some extent at least the curiosity and desire to know more which is such an encouraging characteristic of the keen naturalist of today.

Ernest Neal
Bedford

Preface

Man has had a long history of association with the deer family — economic, religious, aesthetic and social. Deer are widely distributed throughout the north-temperate zones of the world, extending into the tropics in Asia and South and Central America. Throughout this range they were the major food species for man the hunter. As, in Africa, the hunter-gatherer economy relied heavily on exploitation of antelope, so, in other areas, the various deer species provided the main source of meat, hides and other products: sinews for sewing or twine, antlers for picks and other tools, etc. So fundamental were the deer to the subsistence of these peoples, so deeply interwoven with their whole life style, that they were endowed with mystical, magical properties, and became an integral part not only of man's secular existence but of his spiritual world as well. In extreme cases the relationship became so intimate, and human populations became so dependent on one particular species — as the Lapp reindeer-herders of the far north of Scandinavia — that the situation developed almost into one of complete social parasitism: one organised population relying entirely upon one other.

Deer were used in these early days for meat, hides, medicines and mysticism — and so remain in areas where the economy is still based upon subsistence hunting. In other areas, however, as human populations advanced in sophistication, became more reliant on agriculture than hunting and acquired from the Middle East new forms of domesticated livestock — cattle, sheep, goats — so the degree of dependence upon deer diminished. In some areas the close relationships persisted: where the social and cultural traditions were so intermingled with the deer, or where environmental conditions were such that conventional domestic stock could not survive, or fared but poorly, the reliance on deer continued, but developed and adopted the new ideas of an agricultural economy. So the Lapps domesticated their reindeer and, in other parts of the world, other peoples in turn brought their native deer into at least semi-domestication. But such were the exception rather than the rule, and in the majority of cases, as agricultural expertise began to spread, close reliance upon the deer declined. An interest in deer remained, however, and European Man, while no longer dependent upon hunting for food, still continued to hunt for sport and recreation. The meat value of the carcase became less important than the antlers, as trophies, and the pageantry of the hunt itself. Vension, while no longer a staple meat, became a luxury item to be enjoyed. In the East, the additional use of deer products for medicinal purposes remained of tremendous cultural significance. Deer were introduced into special hunting preserves or deer parks where they could be pursued in chase or harvested for venison or other products; and, as man traded further and further afield, or migrated to colonise new lands, so deer were transported from country to country as articles of trade, or established in the 'new lands' for the homesick colonists to hunt. Such translocations and introductions were repeated throughout history and continue right up to the present day. Indeed, it is probably fair to say that Man has had a greater part in the current worldwide distribution of the various deer species than in the distribution of any other non-domestic animal.

Man's influence on the distribution of deer, however, is not merely reflected in an increase in the range of many species through introductions. The effects of hunting and continued habitat erosion as human populations sequestered more and more land for agricultural, silvicultural or industrial development have also come to threaten many vulnerable species with extinction, particularly in more modern times, and much recent interest in the biology of deer had been directed towards conservation.

Since prehistoric time, then, Man has always had a close relationship with deer, manipulating their populations (for exploitation, for conservation — or for control, when the deer threaten his agricultural crops or timber) and influencing their worldwide distribution. While relationships in the developed world have maybe escaped from that fundamental reliance of earlier times, something of the mystical quality they were accorded during that long history of close dependence perhaps lingers on in our modern consciousness. For still, even among the most urbane, there remains an interest in deer, an appreciation of their aesthetic appeal — and still, for most people, the fleeting sight of a wild deer sets the pulses quickening.

This book is an attempt to review something of what we know of the biology and natural history of the world's deer. It cannot hope to be exhaustive, nor would it be appropriate to compile individual accounts of the 40 or so species of deer currently recognised in the world. Other texts (e.g. Whitehead's 1972 *Deer of the World*) have summarised what we know of the different species within such a format. This book will adopt instead a more functional approach, dealing with more general biological questions — about use of habitat, food and feeding, social organisation and reproduction — weaving in information about individual particular species when appropriate within such a comparative context. The book is, as others in the series, unashamedly directed at the enthusiastic naturalist, not the professional scientist. While I may hope that the professional will also find it a refreshing review of current knowledge and ideas, it is not intended as an academic work nor is it an exhaustive compilation of our knowledge of the biology of every species. Rather, its aim is to introduce the non-specialist to the diversity of deer as a group and to attempt to explain something of the tremendous variability that they show in behaviour, ecology and physiology in relation to the range of different environments they have come to colonise.

1 The origins of deer

Within the world of animals the greatest diversity of form and richness of species is to be found among the herbivores — those animals which feed upon plants. This is perhaps not so surprising, for green plants, the primary producers of any ecosystem, are at the base of the food chain. Only they can synthesise for themselves, by exploiting the energy of the sunlight, the complex organic molecules of life; all other life forms must take in those compounds ready formed and thus depend ultimately on the plants (and a few specialised chemo-synthetic producers) for their nutrient intake. Herbivores feed directly upon the plants themselves; carnivores feed upon those herbivores or on other carnivores. So, with herbivory at the very base of all such food chains, selection has favoured the evolution of a range of herbivores from among most animal groups.

During the Mesozoic era (150–250 million years ago) — the Age of Reptiles — vertebrate life on Earth was dominated by the huge diversity of dinosaurs and other large reptiles. Although the first true mammals appeared in the Jurassic period (190–195 million years ago), they remained small and relatively insignificant throughout the remaining 130 million years of the reptile era. Most were shrew-like, and many were probably nocturnal. Only after the extinction of the large terrestrial dinosaurs at the end of the Cretaceous (65 million years ago) did mammals really begin to diversify. Of course, the decline of the dinosaurs at that time suddenly created, so to speak, a 'vacancy' on land for a large vertebrate herbivore: within the Eocene period of geological time, some 50 million years ago, appeared the first mammalian attempts at filling that vacant niche. We find in the fossil record of that time a number of archaic ungulate groups which relatively rapidly — and perhaps, as A.S. Romer suggests, rather prematurely — attained large size and ponderous build, only to face early extinction in competition with more progressive groups. Such forms included the Pantodonts, Dinocerata, Xenungulata — all extinct today; as well as the various sub-ungulates — the elephants and their various relatives — which have persisted to the present day but are represented by only a few, essentially relict, groups. Among the placental mammals of that time two main groups alone of large herbivores, the perissodactyls and artiodactyls, progressed steadily and rather conservatively to the present time.

Modern ungulates are an assemblage of large, hoofed mammals which encompasses almost all the large herbivorous mammals of the world. The group is, however, a somewhat artificial one, for the ungulates, as we have seen, represent the end point of a series of early attempts at large herbivores and consists of a number of distinct and only distantly related orders: the Perissodactyla, the Artiodactyla, and various relict groups such as the Proboscoidea (elephants and hyraxes) and Sirenia (manatees and dugongs). Nevertheless, there are several features in common among the orders. Associated with their herbivorous diet, all ungulates have developed large, flat-topped cheek teeth for grinding tough vegetable materials. Of the two major orders (the Perissodactyla and Artiodactyla) at least, most are fast-running forms, and towards this end the limbs are long and powerful; the toes are lifted only until the tips touch the ground (digitigrade) and the number of toes in the foot is reduced.

Among this modern ungulate assemblage, the most successful orders (and, interestingly, the ones with the greatest similarity, suggesting that their success relates at least in some degree to those common characters they have independently acquired) are, as we have noted, the Perissodactyla (the odd-toed ungulates) and the Artiodactyla (the even-toed ungulates). The perissodactyls, once a much more abundant order, are now represented by only three main groups: the horses, the tapirs and the rhinoceroses. It is among the artiodactyls that we find the widest diversity of living ungulates, and this is the order to which belong the world's deer.

The classification and inter-relationships of the Artiodactyla are relatively complicated. The living forms fall easily into three major groups: (i) the suborder Suina, which contains the hippopotamus, the pigs and wart-hog of the Old World, and the New World peccaries; (ii) the suborder Tylopoda: the camels, llama, vicuña and guanaco; (iii) the suborder Ruminanta or Pecora: the largest suborder, containing the chevrotains (mouse deer), the giraffes and okapi, all the varied bovids (cattle, sheep, goats, antelope and gazelle) and the true deer. While the classification of *living* groups is simple, however, when fossil forms are included such an easy separation becomes clouded by a whole host of families which are intermediate between existing types or belong to extinct side branches, and it is far from clear how the three modern suborders are in fact related to each other (although it is always presumed that the swine and the tylopods are more primitive than the true ruminants).

Even within each suborder relationships are uncertain. Among the Pecora, the earliest forms in the fossil record were small, rather primitive animals, remarkably similar to the living tragulids or chevrotains. The chevrotains are cud-chewers, but the associated subdivision of the stomach is less complex than in other living pecorans (and is at a level not dissimilar to that found among the camel group). Unlike the more 'advanced' pecorans such as the antelope or deer, chevrotains have no horns or antlers. Instead the upper canines are exaggerated (particularly in the males) and developed as weapons. The limbs are long, but, although functionally two-toed, retain four digits; the lateral toes are short and slender, but are complete and not reduced to the splints of the more 'advanced' groups. Finally, while in (presumed) later groups the middle bones of the 'palm' of the hand and foot (metacarpals and metatarsals) are fused in both front legs and hindlegs into a single 'cannon' bone, among the chevrotains this fusion is incomplete in the forelimbs. Such features as this suggest that living chevrotains represent a fairly early stage of development within the Pecora, and suggest that they may have been a side branch of a major, now extinct, group of *traguloids* which later gave rise to the giraffes, bovids and true deer.

These 'more advanced' ruminants appear in the Miocene, and even the earliest fossils show a distinct separation into two groups: the line leading to the modern bovids, and that developing later to the giraffes and deer. The 'cervoids' (deer-like group), like the chevrotains, were all browsing types with low-crowned teeth, typically found in forest or bush habitats. The 'bovoids' by contrast had branched out and colonised open grassland, developing high-crowned teeth more suitable for the tough and highly silicious forage that grass provides. This separation of the two lines was clearly a critical one: the specialisation of each to a distinct habitat and diet clearly underlies many of the differences in both anatomical and physiological characteristics between the two groups, but it also of course permitted each a wider radiation, the development of a greater diversity of species within their own discrete environmental type. A similar 'demarcation agreement' seems to have permitted the separate development within the cervoid group of giraffes and deer. Giraffes remained restricted to the more tropical areas of the world, while the deer colonised and spread through the forests of the north-temperate zone. The first true deer, all relatively small forms, appeared in Eurasia during the Miocene and early Pliocene (approximately 20 million years ago). By the end of the Pliocene there were present in the fossil record a great variety of types, many of which have survived into modern times.

Deer then appear to have developed as a north-temperate group of artiodactyls retaining the forest- or woodland-dwelling habit of their chevrotain-like ancestors. Many of the common features of their physiology and general biology can be related to this origin, while differences between the living species in structure, diet and social organisation reflect more recent variations from the common theme. Indeed, the 40 or so species of deer living today have diversified to occupy such a wide range of different environments that patterns of habitat use, diet, social organisation and behaviour are highly variable, as each has over time become adapted to the particular environment in which that species now finds itself. Thus there is in practice marked variation between species — and not only between species, but between different populations of even the same species when they occur in different environments.

Equally, adaptation to the requirements of the environment and life style are so precise, the pressure of being well adapted to ecological circumstance so strong, that particular characteristics of behaviour, ecology or physiology observed sometimes appear to be more characteristics of the type of environment occupied than of particular species — such that all species or populations adopting a particular life style within a particular environment must evolve the same characteristics. This close adaptation to environmental circumstance and the flexibility within and between species in adapting to new circumstances, new environments, will be a key theme in our discussions through the next few chapters of the natural history of deer.

GENERAL BIOLOGY OF THE CERVIDS AND RELATED ARTIODACTYLS

Characteristic of all the artiodactyls are a number of common features of morphology and physiology, reflecting in part a common ancestry and in part common adaptation to a herbivorous life style — coping with a vegetable diet which is difficult to digest, and avoiding predators. Many of these characteristcs are obvious, the similarities unconsciously recognised: after all, artiodactyls (and particularly the main pecoran group) actúally look the same, and you immediately know when you see one that that is what it is. Most are relatively large-bodied animals with acute senses: large eyes, sensitive ears with large and mobile pinnae, and an acute sense of smell. All have long limbs and strong muscles, enabling rapid escape from predators; in all, the limbs have been further modified for fast running by changing the relative lengths of the various bones, by progressively changing the orientation of the bones of the foot so that the weight is increasingly supported on the tips of the digits, and by reduction of the absolute number of toes. In fact, in all species the number of functional toes in contact with the ground is reduced to two on each foot, but pigs and peccaries, like the more primitive chevrotains (above, page 2), in fact retain the complete structure of four toes on each limb; in higher families these lateral toes are reduced to splints. The modifications of the limb bones are also progressive, and further developed among the bovids and cervoids than in other groups, but in essence the same changes are apparent in all groups (Figure 1.1). The proximal bones

Figure 1.1 Reduction of the digits and metapodials in the artiodactyl limb

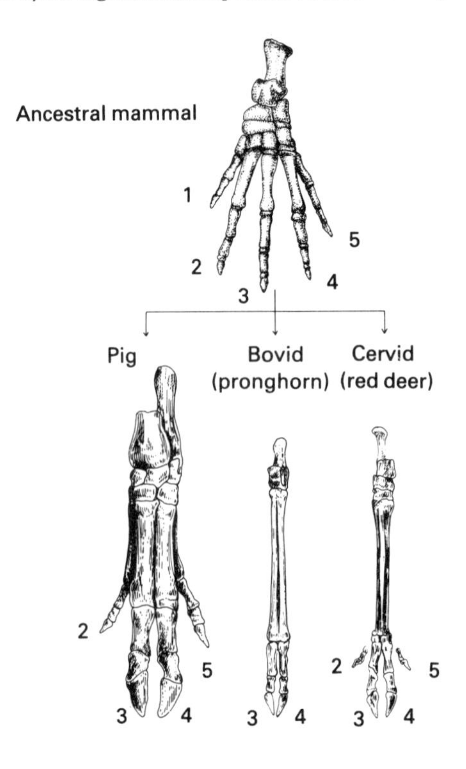

(femur and humerus) are short and stout, resulting in a powerful drive; the second joint (tibia/fibula, radius/ulna) is longer, making for a long fore and aft swing; and the bones of the palm and sole (the metapodials) are greatly elongated, giving a third segment to the limb. As we have noted, there is also in more advanced groups a tendency for fusion of the middle metapodials into a cannon bone.

Although deer, like other large herbivores, are highly adapted in this way, through modifications of the skeleton and muscle systems, for rapid flight in escaping predators, anti-predator tactics are in fact based in the first instance on an ability to detect potential danger well in advance. Thus, as previously mentioned, the senses are acute. The sense of hearing is extremely well developed. Sense of smell, too, is highly sophisticated (although in this case the sensitivity is developed only partially for detection of potential predators: deer themselves possess a variety of different scent glands (page 29) and much intraspecific communication is through scent and scent-marking). By comparison, eyesight, in terms of an accurate pictorial representation of the world around them, does not appear to be strong, and in that sense at least deer are not primarily 'visual' animals.

Neurophysiological studies on the fallow deer, for example, suggest that it can just about make out vague shapes and forms up to a distance of about 50–60m. It is not entirely colour blind (as many have long believed); at close range it has crude and partial colour vision so that the shapes it *is* able to see are perceived in colour. Beyond 60m, however, it is unable to distinguish colour at all, and indeed its ability to 'picture' objects at distances greater than this is virtually non-existent (Allen, 1983). Although this poor visual resolution may seem odd in a relatively advanced mammal, we should remember that over evolutionary time any animal species will develop adaptations only in response to perceived selection pressure. Thus, while deer may not be able to perceive 'camera-sharp' images of the world around them, as herbivores they do not in fact have the same need for acute vision of this kind as might a predator. Instead, as we have noted, one of the strongest selection pressures they do suffer is the risk of themselves succumbing to predation. Just as senses of smell and hearing are highly tuned to provide an early-warning system for potential predators, this too is the main importance, to the deer, of its eyesight. While it may not be able to detect shape or form with any accuracy beyond some 60m, the sensivity of the eyes to *movement* is exceptionally acute, capable of detecting the slightest movement at distances of up to 200–300m; in practice, such ability is of far greater value to the deer than the recognition of precise images.

This development of long limbs and strong muscles for flight, and the highly acute sense organs both represent adaptations to enable the very vulnerable deer, or any other artiodactyl for that matter, to escape from predators. Other adaptations are related more to its herbivorous diet. Characteristic of all artiodactyls apart from the pig group is the presence of a complex stomach generally with three or four chambers. These animals are cud-chewers, and food when first ingested is taken into the first chamber of the stomach and stored. Later it is returned to the mouth and chewed properly before passing to the further stomach compartments.

This method of feeding has two great advantages. Firstly, vegetable material, on which all ruminants feed, is relatively indigestible stuff of generally rather low nutritional quality. Plant cells are surrounded by a thick cell wall composed of cellulose. This gives considerable strength to the cell, making it difficult to rupture to extract the cell contents. More importantly, however, a very significant proportion of the energy value of plant material is actually bound up within the cellulose itself. To those herbivores which are unable to digest cellulose (for example, insects such as locusts and grasshoppers, or caterpillars) and can exploit only the actual cell contents, only some 20 per cent of the energy value of the plant material ingested is actually available; to animals which can digest the cellulose as well, up to 50 per cent or even 60 per cent of the dry matter may be available from the same plants. These figures clearly represent a maximum limit, and the actual digestibility in practice varies considerably from plant to plant. Nonetheless, the ability to digest cellulose at all is clearly at a premium; yet few organisms above the level of very simple creatures such as bacteria, fungi or protists are in fact able to do so. A few animals, such as platyhelminth flatworms or gastropod molluscs (slugs and snails), possess the necessary enzymes, but the capacity to digest cellulose for themselves has been lost by all higher organisms, including mammals. All these higher organisms have to rely on culturing populations of bacteria, fungi or other micro-organisms to do it for them; and the development of specialised gut structures to house these symbionts has been evolved over and over again as, independently, different members of different taxonomic groups have developed a more and more herbivorous specialisation. Termites among the insects, rabbits and hares, sirenians, kangaroos and other marsupial herbivores, and, even among the primates, a few species of highly specialised leaf-eating monkeys, all have independently evolved modifications in gut structure to accommodate the digestion of cellulose by symbiotic micro-organisms.

The odd-toed and even-toed ungulates have faced and resolved the same problem, too, but in different ways. In the gut of perissodactyls — as in fact also in the rabbits and hares — the caecum, a blind sac opening from the junction between the small and large intestines, has become greatly enlarged (Figure 1.2), and in this structure the cellulose of plant cell walls is digested by the caecal microflora. Such a system has its limitations. The main absorptive region of the mammalian gut, that part whose structure is most adapted for the absorption of the nutrient products of digestion, is the latter part of the small intestine. The food passing through the small intestine of a horse, however, has yet to reach the cellulose-digesting microflora of the caecum, while the caecal contents (whose cellulose is at least partially digested and is mixed with bacteria and fungi, themselves nutritious and digestible) discharge directly into the large intestine, whose absorptive power is poor.

The products of carbohydrate fermentation by the micro-organisms of the caecum are mostly a series of volatile fatty acids, such as acetic acid, propionic acid and butyric acid, rather than the more familiar disaccharide and monosaccharide sugars of our carbohydrate digestive process. These volatile fatty acids can be absorbed directly through the caecum wall into the blood stream, while the products of digestion of protein, fats and simpler carbohydrates carried out in the normal mammalian way by enzymes within the stomach and duodenum (the first part of the small intestine) can of course be absorbed as usual before the food mass even

Figure 1.2 The digestive system of a non-ruminant ungulate: the horse, showing simple stomach and enlarged caecum

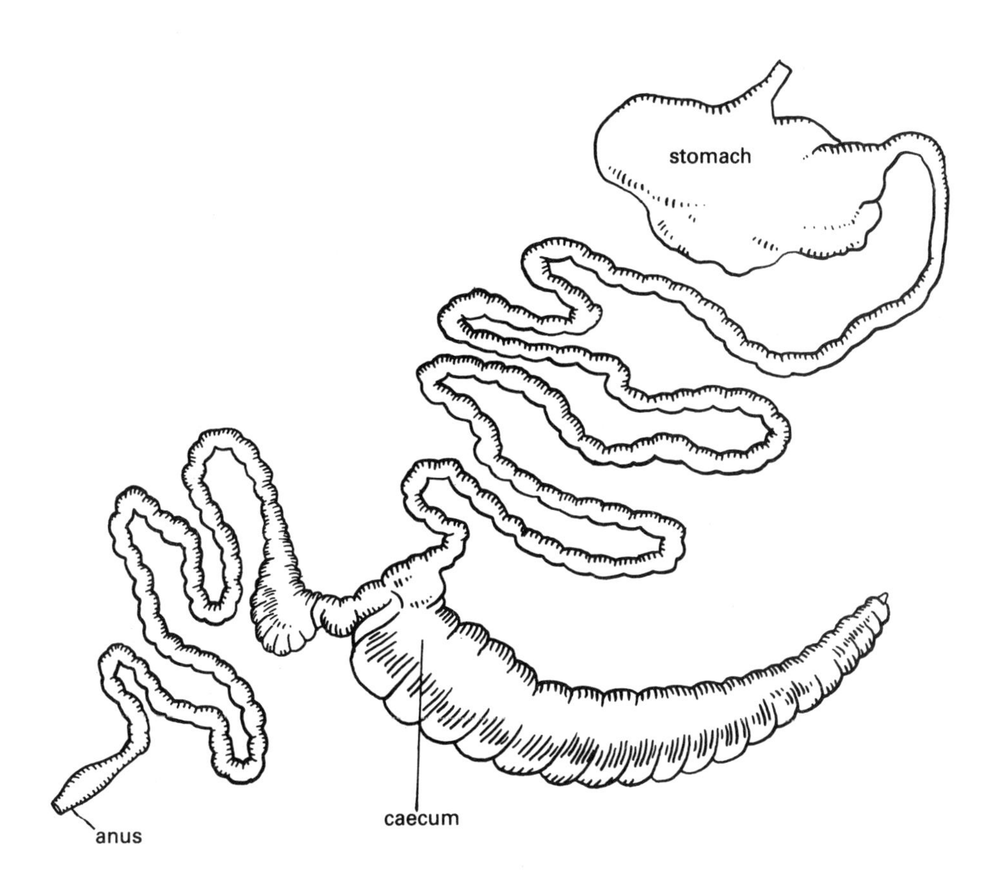

reaches the caecum. The nutritive value of the cellulose itself is not, however, confined just to the volatile fatty acids released by microbial fermentation. The micro-organisms themselves sequester some of the nutrients and energy for their own growth and reproduction; and the caecal contents, as they discharge into the colon, contain large populations of these micro-organisms mixed in with the undigested food remains. The colon produces no digestive enzymes which might break these down; nor does it provide any further absorptive surface for absorption of the products of such digestion, or the continued absorption of digested foodstuffs not absorbed within the caecum: inevitably, some of the potential nutritive value of the food is lost.

The stomach structure of the more advanced artiodactyls is perhaps better adapted. Here the culture chamber for cellulose-digesting micro-organisms appears far earlier in the alimentary tract — within the complex multichambered stomach — and the digested cellulose with its adherent micro-organisms can be more fully digested and absorbed by the 'host', because it has still to pass through the digestive regions of the small intestine (the duodenum) and the absorptive ileum. The tylopod

Figure 1.3 The ruminant digestive system. Inset: detail showing the structure of the rumen complex itself. The route of ingested material is indicated by the arrows. Food is swallowed, and passes from the oesophagus into the rumen, where mechanical breakdown and attack by micro-organisms commences. The partially digested material is repeatedly regurgitated for further chewing. When particle size is sufficiently reduced, food material passes on through the reticulum, omasum and abomasum

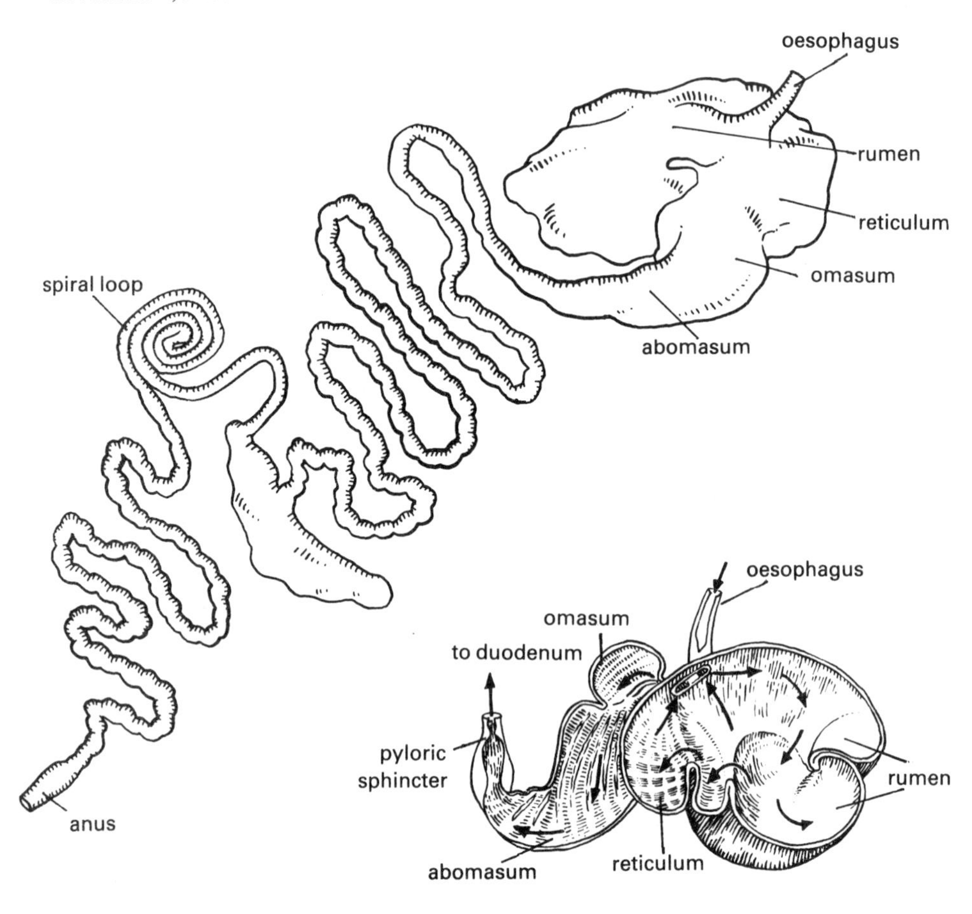

artiodactyls (the camellids) and the most primitive pecorans, the chevrotains, show the beginnings of such developments, but the complex multichambered stomach reaches its fullest refinement within the more advanced ruminants. Here, a complex four-chambered organ is found consisting of rumen, reticulum, omasum and abomasum (Figure 1.3). The omasum and abomasum correspond approximately to the true stomach of monogastric mammals, while the rumen and reticulum are specialised developments of the area where the oesophagus meets this stomach.

The cellulose-digesting micro-organisms are housed within the first two chambers of this complex organ. Vegetation ingested passes directly to the rumen. The partially chewed plant material is mixed with the micro-organisms, and then returned to the mouth, to be rechewed. This regurgitation and rechewing of the cud serves a number of functions: it increases the surface area of the food for bacterial action, the micro-organisms are themselves better mixed into the food medium and, finally, regurgitation of course increases the overall effective length of the first part of the gut, so that the cellulose-digesters have longer to work. The cycle of regurgitation, chewing and reswallowing and the passing of food between rumen and mouth continues over and over again, until particle size within the food is sufficiently reduced that it may pass into the next chamber of the stomach: this, the reticulum, is essentially a fermentation chamber, where a murky suspension of tiny food particles and micro-organisms is held for a considerable period while the cellulose is digested. (As was described for odd-toed ungulates, the products of this microbial fermentation of the cellulose are a series of volatile fatty acids. Just as these could be absorbed directly across the wall of the perissodactyl caecum, so, in ruminants, they may be absorbed directly into the blood stream from the rumeno-reticulum.) Finally, the mixture of digested cellulose, plant-cell contents and micro-organisms passes on into the omasum and abomasum (where digestion of fats and proteins and other elements of the ingesta is commenced exactly as in the stomach of monogastric mammals) before passing into the small intestine, where the processes of digestion and absorption may be completed in the normal way. Such a process ensures a remarkably efficient exploitation of the plant material on which such ruminants depend.

This description is necessarily a rather general one and it is striking how, even within the ruminant group, the exact structure, development, relative size and importance of the different chambers of the 'stomach' vary between species, reflecting subtle differences in diet. Thus, in those species which feed very choosily, selecting from their general vegetational environment small morsels of relatively higher nutritional value, the reticulum need not be especially well developed, for more of the total nutritional requirements of the animal can be met from cell contents without necessarily digesting fully the cellulose of the plant cell walls. By contrast, unselective bulk feeders, which tend to take in rather larger quantities of poorer-quality forage, must derive more from the cellulose itself. In such species both rumen and reticulum are very considerably larger in proportion. Figure 1.4 illustrates this general point by showing a comparison of the stomachs of a roe deer (as a relatively selective feeder) and an African buffalo (as the epitome of a bulk digester). The true extent of variation, however, is delightfully subtle and will be examined in more detail among the Cervidae in Chapter 3.

We noted on page 6 two great advantages of the ruminant stomach. The second advantage is linked again with a vegetable diet. Since, even with the help of ruminal micro-organisms, the digestibility of plant material is relatively low, all herbivores must make up for this by taking in relatively large quantities of forage. When feeding concentratedly, however, they are of course at considerable risk from predators, for their attention is necessarily rather divided. The four-chambered structure of the ruminant stomach allows them to overcome this problem: by using the expanded volume of the rumen chamber much as a rodent uses cheek pouches, or a bird its crop. They feed very rapidly, not stopping to chew their food, but cramming it into the rumen for storage. When the rumen is full, they may retire to a safe place to regurgitate the material and chew it properly. The stomach structure of the ruminant thus solves simultaneously the problems of reducing exposure to predators while feeding and of digesting its relatively indigestible plant food. And of

Figure 1.4 Relative proportions of different parts of the gut in a bulk feeder (a: African buffalo) and a more selective forager (b: roe deer). Source: Hofmann, 1985

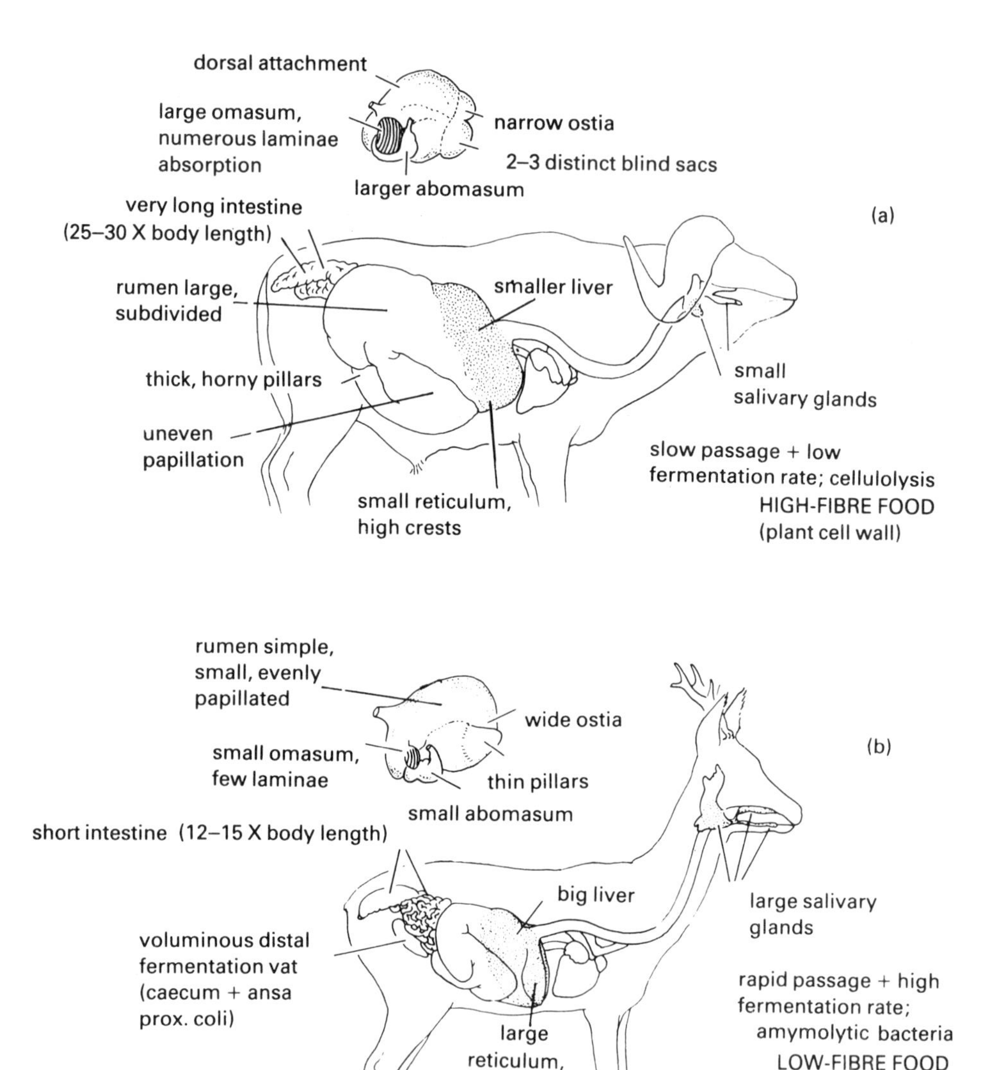

course the two things are far from independent. The ruminant is more exposed to predation because of its indigestible diet and the resultant need to take in large quantities of food; likewise, it is actually the distension of the rumen, registered by sense organs in the muscle wall, that tells the animal to stop feeding and automatically triggers the regurgitation of the rumen contents for cudding.

HORNS AND ANTLERS

One other characteristic of the more advanced groups among the ruminant artiodactyls is the possession of horns or antlers, bony structures supported upon the skull. True horns are a bovid speciality: single unbranched sweeps of bone arising from the skull of, most commonly, both sexes, growing throughout life and covered by an outer sheath of keratin or horn. Antlers are a cervid characteristic, or to be more precise are found only among deer; for, while it is true that only deer ever have antlers, not all species of deer do develop them, as we shall see. Antlers, where present, are also borne in pairs and are again structures of solid bone, but here the similarity to horns stops. Antlers have no outer covering when fully formed; they do not grow directly from the skull itself, but are supported on basal discs of bone, or 'pedicles' (Figure 7.1); and they are usually branched. Most importantly, while horns are permanent structures retained throughout life, the antlers of deer are shed every year and regrown. The new, and usually more complex antler of the following year regrows from the pedicle, surrounded during its development by a thin skin, or 'velvet', richly supplied with blood capillaries. When the antler is fully formed the blood capillaries constrict at the antler base, and the velvet shrivels and falls off, leaving the newly-developed antler exposed. Finally, while horns are generally borne by bovids of both sexes (though not exclusively: in some species they are found on males only), antlers are usually developed by male deer alone; only in reindeer and caribou are antlers borne by both sexes. Despite their superficial similarity, then, horns and antlers are quite distinct structures. Developed independently of each other, they may even serve somewhat different functions. Horns are almost certainly primarily weapons of combat or defence, while antlers, although they are indeed used in aggressive encounters, are probably at least as important in social display (Chapter 7).

Even though antlers are found only among the deer family, not all deer possess such structures, and we thus cannot rely on this unique character to distinguish deer from other ungulates. Indeed, it is evolutionary history rather than any one morphological or physiological character that defines the deer, for morphology, physiology and behaviour of all the species are closely adapted to each one's environment. Although the deer were primarily a woodland group while bovids developed in open grasslands, both families have since diversified into a whole range of overlapping environments; the common environmental pressures of a given habitat have drawn the same convergent adaptations of morphology, physiology and behaviour from whatever ungulate — whether bovid or cervid — has colonised it, and on such characters alone the two groups are not easy to distinguish.

2 Modern deer

THE VARIOUS SPECIES OF DEER

Modern deer are divided into five subfamilies: the Chinese water deer (Hydropotinae); the muntjacs of India and Asia (Muntiacinae); the main Eurasian group, Cervinae, which (perhaps not surprisingly, since it occurs in that geographical area where deer first evolved and diversified) contains the greatest number of genera and species; the Odocoilinae, a group containing the roe deer, moose and reindeer, but otherwise represented primarily by the New World deer; and the Moschinae. This last subfamily is recognised only by those who include the little musk deer of the Himalayas, China and southeast Asia within the cervid family at all. Many authorities consider that it should be separated into a distinct family, the Moschidae; but, even if not a true cervid, it is clearly closely related to the other deer and for completeness it will be included in our discussions in these pages.

Table 2.1 lists the various deer species according to these subfamilies, showing for each its geographic range and extent of variation over that range in terms of the number of subspecies recognised.

The legacy of the Eurasian origins of the deer is very clear. Many groups are still Asian or Eurasian in distribution; these same groups contain within themselves the highest number of species and races. The present-day distribution of many species has been influenced by translocations by man (Chapter 8); if such translocations are disregarded, however, the majority of all deer in the Old World are seen to be restricted to the northern hemisphere (the only exception is the little Bawean deer, endemic to the Bawean Island in the Java Sea: page 22). Only in the New World are deer found more generally in the southern hemisphere, with ten species of Odocoilines, distributed across five genera, occurring in South America. Most species also retain a primarily woodland habit, although some of the smaller Eurasian species (roe deer, hog deer) have adapted readily to more open environments, as have the pampas deer of South America, and other species such as the diminutive Chinese water deer, the barasingha and the thamin have become specialised to live in more marshy areas.

Table 2.1 Modern deer: Family Cervidae

Subfamily Moschinae : Musk deer

Moschus moschiferus : Siberian musk deer
(USSR, Korea, northern China) 3 subspecies
Moschus berezovskii : Dwarf musk deer
(Southern China and Vietnam)
Moschus chrysogaster : Himalayan or alpine musk deer
(West and southern China, Burma, Kashmir, India and Nepal) 5 subspecies

Subfamily Hydropotinae : Water deer

Hydropotes inermis : Chinese water deer
(China, Korea; introduced UK and elsewhere) 2 subspecies

Subfamily Muntiacinae : Muntjacs and tufted deer

Muntiacus muntjac : Indian muntjac
(India, Sri Lanka, Tibet, southwest China, Burma, Thailand, Vietnam, Malaysia and Indonesia) 15 subspecies
Muntiacus reevesi : Reeves's or Chinese muntjac
(East China, Formosa; introduced UK) 2 subspecies
Muntiacus crinifrons : Hairy-fronted muntjac or black muntjac
(Eastern China)
Muntiacus feae : Fea's muntjac
(Thailand, Tenasserim)
Muntiacus rooseveltorum : Roosevelt's muntjac
(North Vietnam)
Elaphodus cephalophus : Tufted deer
(Burma, southern and central China) 3 subspecies

Subfamily Cervinae : Eurasian deer

Cervus elaphus : Red deer
(Europe, north to Scandinavia; North Africa (Barbary deer), Asia Minor, Tibet (shou), Kashmir (Kashmir deer or hangul), Turkestan (Yarkand deer), and Afghanistan (Bactrian or Bokharan deer); introduced Australia, New Zealand) 12 subspecies
Cervus nippon : Sika deer
(Japan, Formosa, Vietnam, Manchuria, Korea, China; introduced UK and New Zealand) 13 subspecies
Cervus canadensis : Wapiti
(Western North America, eastern China, Manchuria, Mongolia; introduced to New Zealand) 13 subspecies
Cervus unicolor : Sambar
(Philippines, Indonesia, south China, Burma, India and Sri Lanka; introduced Australia and New Zealand) 16 subspecies
Cervus timorensis : Rusa or Javan deer
(Indonesia; introduced Australia, New Zealand, Fiji, New Guinea) 6 subspecies
Cervus duvauceli : Swamp deer or barasingha
(Central and northern India, Nepal) 2 subspecies
Cervus eldi : Eld's deer, thamin, or brow-antlered deer
(Manipur, Thailand, Vietnam, Burma, Tenasserim) 3 subspecies
Cervus albirostris : Thorold's deer
(Tibet and China)
Elaphurus davidiensis : Père David's deer
(Not known in the wild state; formerly throughout China, Korea, Japan, now being reintroduced into China)

Table 2.1 Continued

Dama dama : Fallow deer
(Widespread Europe and Asia; introduced Australia, New Zealand, Africa North and South America) 2 subspecies
Axis axis : Chital or spotted deer
(India, Sri Lanka; introduced New Zealand) 2 subspecies
Axis porcinus : Hog deer
(India, Sri Lanka, Burma, Thailand, Vietnam; introduced Australia) 2 subspecies
Axis kuhli : Kuhl's or Bawean deer
(Bawean Island)
Axis calamianensis : Calamian deer
(Calamian Islands)

Subfamily Odocoilinae

Capreolus capreolus : Roe deer
(Europe and Asia, north to Scandinavia, Siberia, east to China and Korea) 3 subspecies
Odocoileus virginianus : White-tailed deer
(North and central US, Canada, northern parts of South America; introduced Scandinavia, New Zealand) 38 subspecies
Odocoileus hemionus : Mule deer
(Western North America, Central America) 11 subspecies
Mazama americana : Red brocket
(Central and South America) 14 subspecies
Mazama gouazoubira : Brown brocket
(Central and South America) 10 subspecies
Mazama rufina : Little red brocket
(Venezuela, Ecuador, southeast Brazil) 2 subspecies
Mazama chunyi : Dwarf brocket
(Bolivia and Peru)
Pudu puda : Southern pudu
(Chile and Argentina)
Pudu mephistopheles : Northern pudu
(Ecuador, Peru, Colombia) 2 subspecies
Blastocerus dichotomus : Marsh deer
(Brazil to Argentina)
Ozotocerus bezoarticus : Pampas deer
(Brazil, Argentina, Paraguay, Bolivia) 3 subspecies
Hippocamelus bisulcus : Chilean huemul
(Chile, Argentina)
Hippocamelus antisensis : Peruvian huemul
(Peru, Ecuador, Bolivia, northern Argentina)
Alces alces : Moose
(Northern Europe, Canada, northeastern US) 6 subspecies
Rangifer tarandus : Reindeer or caribou
(Scandinavia, European Russia, Greenland, Canada, Alaska; introduced arctic and antarctic islands) 9 subspecies

Although in Table 2.1 we listed some 40 species, the classification of deer is complex and controversial; the number of species recognised in each group by different workers depends upon whether they accord certain recognisably distinct geographic races specific or subspecific status. Some authors consider for example many of the different races of sika deer as distinct species (e.g. Ellerman and Morrison-Scott, 1951; Whitehead, 1972); others consider the whole genus *Cervus* (wapiti, red deer, rusa, sika, sambar, barasingha and Eld's deer) as a single aggregation (e.g. Harrington, 1982); while others again suggest that, while the red deer group (Scottish red deer, European red deer and wapiti) and the sika deer of the Japanese mainland (*Cervus nippon nippon*) are distinct species, all other sika, and possibly also rusa deer and sambar, originated as hybrids between the two (e.g. Lowe and Gardiner, 1975; Harrington, 1985). Yet again, while many authors classify musk deer within the family Cervidae (e.g. Flower, 1875; Bell, 1876; Simpson, 1945), others, while agreeing that they show superficial resemblance to other living deer, consider that the various musk deer should be placed in their own distinct family, the Moschidae (Gray, 1821; Brooke, 1878; Flerov, 1952; Groves and Grubb, 1987). Our list here is thus necessarily somewhat arbitrary and oversimplistic.

Before we explore further the inter-relationships between the various species we have mentioned, however, it is perhaps appropriate to offer each a brief introduction. (Obviously, in a work of this kind, it is possible to offer only the scantiest of treatments; references are included at the end of each section to point the reader to sources of more detailed accounts of individual species or groups where appropriate and, for an overview, the reader is also recommended to consult Whitehead, 1972.) Plates 1–20 accompany the text.

Musk Deer

Musk deer, if included in the Cervidae at all, are usually recognised as relatively primitive forms. Small 'deer' of approximately 10–15 kg and standing about 0.5 m to the shoulder, they are somewhat hare-like, with long ears, arched back and bounding gait. Unusually among the deer, females are heavier than males (male 10–12 kg and female 10–15kg for *Moschus moschiferus*: Egorov, 1965). Both sexes lack antlers, but males possess large canine tusks in the upper jaw (this is not an uncommon character among smaller cervids with little or no antler growth, and is shared by both muntjacs and Chinese water deer). These canine tusks, again like those of muntjac and water deer, are movable in their sockets, an adaptation which it has been suggested facilitates feeding and cud-chewing (Cooke and Farrell, 1981; Dansie, 1973). Musk deer, like other advanced pecorans, are ruminants and have a four-chambered stomach, but rumen structure is simpler than that of most cervids. Furthermore, as we have already suggested, musk deer possess a number of other characters which are not shared with other living deer: the liver has a gall-bladder (a bovid feature not found in any other cervid), and musk deer have only one pair of teats (other cervids have two pairs). Finally, musk deer of course take their name from the adult males' curious musk gland, a modified preputial scent gland (see page 160, Chapter 8).

Characteristic of dense montane forests, musk deer are distributed sporadically throughout the mountainous areas of Asia from just north of the Arctic Circle southward to the northern edge of Mongolia and to Korea. Further south, but avoiding the Gobi Desert, they also occur in China, northern Vietnam, Burma, Assam and the Himalayan region (Green, 1985). The taxonomy is complex and controversial: for our purposes, we shall, following Green (1985), recognise three distinct species: *Moschus moschiferus* Linnaeus, the Siberian musk deer of USSR, Korea and northern provinces of China; *M. berezovskii* Flerov, the dwarf musk deer of southern China and Vietnam; and *M. chrysogaster*, the alpine or Himalayan musk deer, of western and southern China, Burma, Kashmir, northern India and Nepal.

(References for musk deer: Flerov, 1952; Green, 1985, 1986, 1987.)

Water Deer

The subfamily Hydropotinae, or water deer, is represented by a single species: *Hydropotes inermis*, the Chinese water deer. These small deer are actually very similar in appearance to musk deer. They, too, stand about 0.5m at the shoulder (male 52 cm, female 48 cm) and weigh between 8 kg and 15 kg — although in this case males are heavier than females (male 11–14 kg, female 8–11 kg). Once again neither sex grows antlers, and, as we have noted, males possess enlarged upper canines forming tusks up to 7 cm in length. Despite this superficial similarity, water deer are certainly true cervids — and, while musk deer live in dense montane forest, the Chinese water deer is a species of lowland grasslands and marshes. Occurring naturally in China and Korea, the water deer has been introduced elsewhere in captivity (as, for example, the UK), where feral populations may become established through the inevitable escapes from parks or zoological collections.

(Reference for water deer: Cooke and Farrell 1981.)

Muntjacs

In Table 2.1 we recognise six species within the muntjac subfamily: five true muntjacs and the related tufted deer. All are once again Asiatic in distribution and, like the musk deer and water deer, the muntjacs are regarded as a relatively primitive group, mainly because of their small size and the fact that, as in those other two subfamilies, males have enlarged upper canines. These canine tusks, however, being only some 2.5 cm in length, are not so well developed as those of water deer or musk deer, and muntjac males also bear simple antlers. These antlers are rarely more than mere spikes, although in mature males of some species a short brow tine may develop. In all species, the main prong is itself short and slender and the pointed tip is curved inwards. Most characteristically, the bony pedicles which support the antler do not originate on the top of the skull as in other deer, but extend some distance down the face, visible externally as prominent ridges (Figure 2.1). Five main species are recognised within the genus Muntiacus, with an additional related species, the tufted deer (*Elaphodus cephalophus*), placed in a genus of its own. All are relatively small deer of woodland or thick forest, ranging in size from 11 kg to 18 kg and with a shoulder height of some 41 cm (Reeves's muntjac) to 57 cm (Indian muntjac) or 63 cm (tufted deer).

The type species, the Indian muntjac (*Muntiacus muntjac*), is widely distributed throughout much of Asia, through India and Sri Lanka into

Figure 2.1 Antler development in muntjac deer (from A Field Guide to British Deer, *ed. F.J. Taylor Page)*

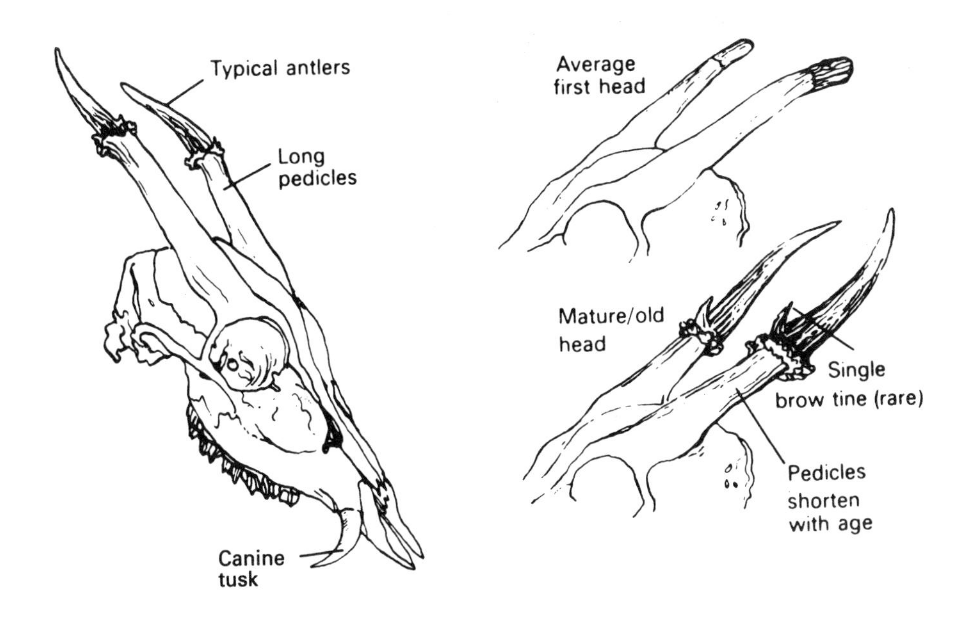

Tibet, southwest China, Burma, Thailand, Malaysia and Indonesia. The Chinese or Reeves's muntjac (*M. reevesi*), a smaller species, is restricted to eastern China and Formosa. Of all the species, these two are the most abundant.

Sympatric with Reeves's muntjac in eastern China is the larger hairy-fronted or black muntjac (*M. crinifrons*), one of the species with secondary branches on the antler, with a little spike developed on the inner side of the base of the main antler. This species, like Fea's muntjac (*M. feae*) of Thailand, is now considered endangered. Fea's muntjac differs from *M. muntjac* primarily in having a darker coat colour and yellow hairs to the centre of the pedicles. Roosevelt's muntjac (*M. rooseveltorum*) of Vietnam is also very similar to the Indian form. (While only five species of true muntjac are generally recognised, Groves and Grubb (1982) suggest that there may in fact be six. They consider that some of the specimens collected by early naturalists in Borneo were erroneously classified with material pertaining to the Bornean subspecies of *Muntiacus muntjac,* yet should have been distinguished as relating to a different species. They propose that two muntjac types occur together on Borneo: the Bornean subspecies of Indian muntjac, *Muntiacus m. rubidus,* already recognised; and a new endemic species, *Muntiacus atherodes*. Confirmation of *M. atherodes* as a true species must, however, await new records.) Finally, the tufted deer (*Elaphodus cephalophus*) of China and northeast Burma is distinguished from the muntjacs by being essentially a montane species, occurring in forests up to 4,570 m. It is also bigger than other muntiacines (63 cm at the shoulder), but has small, unbranched antlers which are often completely hidden by tufts of hair which grow from the forehead.

(References for muntjac: Schaller, 1967; Mishra, 1982; Chapman, 1988.)

Eurasian Deer

The main Eurasian subfamily, the Cervinae, contains perhaps the greatest diversity of all the deer, as well as many of the most familiar species. Fourteen species in four genera are included in our list in Table 2.1. but this actually extends to some 75 distinct subspecies.

Cervus species: The largest and by far the most widespread of the four genera is the genus *Cervus*, which includes the familiar red deer of Europe, the wapiti of North America, the sambar of India and Asia, and the Japanese and Asian sika deer — a group of species which are all generally somewhat similar in both appearance and habit. The genus also embraces a number of other less typical species such as the barasingha or swamp deer of India and Nepal, Thorold's deer from Tibet, and the thamin or Eld's deer of East Asia.

Red deer (*Cervus elaphus*) occur widely through Europe and Asia, from Scotland and parts of England and Ireland, through mainland Europe to North Africa and into Asia. Red deer of British origin have been introduced into Australia and New Zealand. The species varies tremendously (mostly in size and weight) across its range and, although its apparent distribution is wide, quite distinct subspecies occur in most of the different regions of its range. Thus, while within Europe the Scandinavian form, *C. elaphus elaphus*, may be considered typical, the Scottish race has long been regarded as a separate subspecies and the races occurring in western Europe, Spain, Corsica and Sardinia and North Africa are also accorded subspecific status. It is, however, clear that all these subspecies are basically red deer at heart. Some of the other variants are perhaps more distinct (and have by various authors been recognised as separate species): these include the maral (*C. e. maral*) in Turkey and Iran, the Bactrian deer (*C. e. bactrianus*) of northern Afghanistan and Turkestan, the hangul (*C. e. hanglu*) of Kashmir and the shou (*C. e. wallichi*) of eastern Tibet. The taxonomy of the whole genus is confused, and it is far from clear why all these latter forms, many of them really quite distinctive, should be considered merely subspecies of the red deer, while others, such as the rusa deer (*C. timorensis*) of Timor, the thamin (*C. eldi*) or even the sambar (*C. unicolor*), themselves apparently little more distinct, should each be accorded their own full species. We shall return to this point later, but it is perhaps worth noting even at this stage that many of these species and subspecies are defined by geographical isolation. When they are brought into contact, hybridisation is commonplace; and there are even those who argue (e.g. Harrington, 1985; and see page 33) that there is no such thing as a pure-bred red deer left in existence, that all are now of hybrid status.

Red deer are medium to large deer, but there is extensive variation in the different forms. The big, western European *C. e. hippelaphus* may weigh up to 300 kg, while the little island race of Corsica (*C. e. corsicanus*) may be as little as 100 kg. All are relatively coarse-coated animals of a red or brownish-red colour in summer pelage (becoming darker and greyer in winter) with a distinct lemon-yellow rump patch. Males carry strongly-branched antlers which, while only a single spike during their first season, develop increasing complexity from year to year, until in mature animals they have up to 20 terminal points. Still primarily woodland animals, red deer have adapted to more open habitat, as for example in Scotland, where some populations survive entirely upon open moorland. It is notable, however, that such open-ground populations tend to perform less well and are frequently smaller in size.

In the far northeast of Asia and across into North America, the typical red deer is replaced by the wapiti (*C. canadensis*), sometimes referred to in the US as elk. Extremely similar to true red deer in all respects except size (wapiti can reach 160 cm at the shoulder and weigh from 240 kg to 450 kg), *Cervus canadensis* is by some authorities considered only another subspecies of red deer (Ellerman and Morrison-Scott, 1951; Flerov, 1952; Corbet, 1978). Others regard it — usually on the grounds of size, colouring and vocalisations (Clutton-Brock *et al.*, 1982) — as a distinct species and indeed may recognise up to 13 subspecies of wapiti. Within the US, for example, different races are identified in *C. c. roosevelti, C. c. nelsoni, C. c. manitobensis* and *C. c. nannodes*. (Subdivision can, however, continue indefinitely: every ecological population differs subtly in some way from every other and each could perhaps lay claim to subspecific status. One wonders to what extent finer and finer recognition of such subspecies reflects more local patriotism than any real taxonomic significance!)

As we shall see, similar problems surround the taxonomy of sika deer, a small *Cervus* of Japan and neighbouring islands. The mainland Japanese sika (*C. nippon nippon*) is usually recognised as the type for the species. Weighing about 48–50 kg, this is an animal a little shorter than the red deer (shoulder height about 80 cm) and somewhat stockier. The winter coat is similar to that of the red deer, thick and grey-brown, tending even to black in the males; the summer coat, however, is chestnut-brown liberally spotted with white (Plate 5). The antlers are shorter than those of red deer and never as branched (with a maximum of four or five tines, or branches, on each side). The various subspecies of the Japanese islands, Vietnam, Formosa, Manchuria, Korea and southeast China differ primarily in size, and in spottiness, with some races retaining the white spots of summer also in the heavier winter coat. Once more, all are essentially woodland and forest deer.

The sambar (*C. unicolor*) is the 'red deer' of India, distributed in fact from the Philippines, through Indonesia, south China, Formosa and Burma to India and Sri Lanka. Like red deer, wapiti and sika, it has been widely introduced into Australia and New Zealand, where it has established wild populations. This again is a woodland species, a large deer, uniformly dark brown in colour (hence *unicolor*) but with a lightening of the brown under the chin, on the insides of the limbs and under the tail. No fewer than 16 subspecies are recognised, so size and weight are variable (shoulder height 61–142 cm, weight 230–270 kg).

The most widespread species of deer in the Indonesian archipelago is the rusa deer (*C. timorensis*), a species closely similar to the sambar and (here we go again!) formerly considered by some authorities to belong to the same species. Rusa are island deer, and six or seven subspecies are recognised in different islands of the Indonesian archipelago. Rusa have

Figure 2.2 The brow-antlered deer is unusual among the cervids in that, unlike all other, digitigrade, deer, its weight is supported on its pasterns as well as on its toes

been introduced into Australia, New Zealand, Fiji and New Guinea. Like the sambar, they are deer of woodlands and open forest, but where they have been introduced elsewhere they appear relatively flexible in habit and can survive well in more open habitats.

A number of *Cervus* species differ more markedly from this series of 'red deer'. Two swamp-dwelling species, the swamp deer or barasingha (*C. duvauceli*) of India and Nepal and the brow-antlered deer, thamin or Eld's deer (*C. eldi*) of Burma and Thailand, both show clear adaptation to this rather different environment. Eld's deer (also called 'brow-antlered deer' because the long first branch of the antler — the brow-tine — and the main beam form a continuous bow-shaped curve) is another red-brown deer of around 110–120 cm in height, but it has whitish underparts, and white on the chin, around the eyes and on the margins of the ears. A peculiarity of *C. eldi* is that the foot has been modified to enable the deer to walk upon the marshy ground of its principal habitats. Rather than being strictly digitigrade and walking on the 'finger-tips', as do most artiodactyls, Eld's deer moves with the weight also supported by the undersides of the hardened pasterns (Figure 2.2). Strictly speaking, the various common names used here (Eld's deer, brow-antlered deer and thamin) are not synonymous: Eld's deer and thamin are used to refer to distinct subspecies: *C. e. eldi* from Manipur, and *C. e. thamin* from Tenasserim in Burma and adjacent parts of Thailand. A third subspecies *C. e. siamensis* extends from Thailand into Vietnam and Hainan Island. All three subspecies are rare, with numbers of the Manipur form estimated in 1976 at only 18 individuals (Ranjitsingh, 1978).

The two races of the swamp deer or barasingha (*C. duvauceli*) are also listed as endangered. Although, during the last century, swamp deer were recorded all along the base of the Himalayas, from upper Assam to Bhawalpur and Rohri in the upper Sind, today viable populations exist only within three national parks in India and in one reserve in Nepal (Singh, 1978). This is an animal of 120–125 cm at the shoulder and weighing 170–180 kg, living in extensive areas of marshy grassland. The coat is brown to yellowish-brown and most distinctive, for the hair is extremely fine, often even woolly, in texture, with tufts of wool also sprouting inside the ears. The neck of the male is maned; the antlers are large and usually bear eight to ten branches on each side.

Finally within the genus *Cervus* is included Thorold's deer (*C. albirostris*), a curious species of high-altitude forest, rhododendron and alpine grassland in eastern Tibet and China. While believed rare, its actual status is unknown: most recent estimates are of some 10,000 individuals within the Quighai province of China. This is a deep brown cervid of about 140 kg, with a creamy belly, white nose and lips, white on the chin and throat and a white patch near the ears (cf. *C. eldi*). Its most distinguishing character is a reversal of hair direction on the withers, giving the appearance of a distinct hump. Further, the hooves are high, short and wide, like those of cattle (Whitehead, 1972).

One other species which must be mentioned here is the curious Père David's deer (*Elaphurus davidiensis*). Although currently separated into a distinct genus of its own, it clearly shows close relationships to the red deer assembly. Discovered in 1865 in the Imperial Hunting Park of Nan-Hai-Tze just south of Peking by the French missionary and explorer Armand David, this curious animal has never been known outside deer parks and zoos. Although originally of fairly widespread distribution throughout China, Korea and even Japan, it appears that it became extinct in the wild some two or three thousand years ago, surviving only in the Imperial Park in Peking. From that single population, specimens reached various zoos and private collections in Europe; and these latter provide the sole source of current populations, for the Nan-Hai-Tze herd was destroyed by 1900. European populations, particularly those of the Duke of Bedford's estate at Woburn in England, have steadily increased in numbers, however, and recently two separate reintroductions of Père David's deer have been made into China (see page 154).

Père David's is a large deer of some 120–130 cm at the shoulder and weighing up to 200 kg. Most strikingly, the antlers of the male appear to be placed back to front. In all other deer with branched antlers, the main beam is erect or backward-sloping from the skull, with all branching tines projecting forwards from that beam. In Père David's deer, all tines project to the rear. Another curious characteristic is the possession of very large, wide-splayed hooves, suggesting that the species was adapted to travel over very soft ground of perhaps snow.

Dama and Axis deer: The Eurasian subfamily Cervinae contains representatives of two other genera: *Axis* and *Dama*. The genus *Dama* is represented by a single living species, the fallow deer, of which two subspecies are recognised: the European fallow (*D. dama dama*) and the Persian fallow (*D. d. mesopotamica*). Persian fallow are close to extinction in the wild, with perhaps 100–120 animals established in three wildlife refuges in Iran. European fallow are far more widely distributed. Although believed to be of Mediterranean origin, fallow were widespread throughout Europe some 100,000 years ago. They probably became extinct in the last glaciation except for a few small refuges in southern Europe. Their second radiation from these relict populations to the rest of Europe, and their introduction to the British Isles, appear to have been at least assisted by Man. Indeed, of all the deer, fallow are perhaps the species whose current distribution has been most influenced by Man; further importations have taken this species well beyond Europe and it is now established in the wild in both North and South America, Africa and Australasia, as well as in its native Eurasia. Both European and Persian fallow are medium-sized deer (females 70–80 cm at the shoulder, males larger, up to 90 cm; weight 35–60 kg for females, 70–100 kg for males). The coat in summer is typically fawn, with white spots on the back and flanks; in winter greyish-brown, with the spots less clearly discernible. As we have just noted, however, over much of their range European fallow owe their origin to human introductions — deliberate, or as escapes from captive herds. As a result of these captive origins, and perhaps many generations of captive breeding, the form is highly variable and there are many colour varieties, from pure white through to melanistic (Chapman and Chapman, 1975; Chapman and Putman, 1988). The most distinguishing feature of the fallow deer, however, is the broadly palmate antlers of the male (Plate 8).

Fallow deer would appear to be fairly closely related to the axis, chital or spotted deer of India and Sri Lanka, the first member of our final Eurasian genus, *Axis*. Of similar size to fallow deer, and like that species an inhabitant of open woodland and forest edge, often foraging out into more open country, the chital (*A. axis*) is, as one of its alternative names suggests, also a spotted deer. The coat is rufous-fawn with white spots on the back, these spots persisting in both winter and summer coat. The antlers, however, are not palmate but curve out and back in a striking lyre shape.

The hog deer (*A. porcinus*) is a much smaller animal, and a species that has left the woodland for tall grasslands or 'grass jungles'. The typical form, *A. p. porcinus*, occurs throughout Sri Lanka, the grass plains of the Indus and Ganges valleys, through Sind, the Punjab, Bengal and Assam into Burma. Further south in Thailand a slightly larger race, *A. p. annamiticus,* replaces it and spreads eastwards into Vietnam. Hog deer males stand about 65–75 cm at the shoulder (with females slightly smaller at 60–65 cm) and weigh 35–45 kg. They are stocky, heavily-built animals, and it is for this appearance, together with their characteristic of rushing through long grass in the manner of a wild pig rather than bounding over it, that they have been called the hog deer. The two other *Axis* species closely resemble *A. porcinus*: Kuhl's or Bawean deer (*A. kuhli*), a species restricted to Bawean Island in the Java Sea between Java and Borneo; and the Calamian deer (*A. calamianensis*), the only deer apart from the sambar found in the Philippine group of islands and itself restricted to the Calamian group. Both species are, not unnaturally in view of their restricted distributions, rather rare, with numbers of *A. kuhli* estimated at 200–500 and of *A. calamianensis* at about 900 individuals in 1977 (Cowan and Holloway, 1978).

Clearly the Eurasian subfamily, the Cervinae, is a much more diverse and varied group than others we have considered to date, and no single reference adequately covers all species. References quoted here for further detail thus relate to single species or to species groups, but the reader is also referred to Whitehead (1972) for a more general treatment.

References: Red deer: Mitchell *et al.,* 1977; Clutton-Brock *et al.,* 1982. Sika: Horwood and Masters, 1970; Ratcliffe, 1987; Takatsuki, 1987. Wapiti: Boyce and Hayden-Wing, 1979; Thomas and Toweill 1982. Sambar: Schaller, 1967. Barasingha: Martin, 1978; Schaaf, 1978. Chital: Schaller, 1967; Eisenberg and Lockhart, 1972; Mishra, 1982. Hog deer: Prater, 1934; Mishra, 1982. Fallow: Chapman and Chapman, 1975, 1980; Chapman and Putman, 1988.

Odocoilinae

Our final subfamily of deer also has European representatives, but the majority of these have distributions to the east which extend across into North America, and many Odocoilines are exclusively New World species. One species alone — an ancient species which is perhaps close to the line from which the others arose — does not occur in the New World. This is the European roe deer (*Capreolus capreolus*), a tremendously successful little deer and an opportunist which, although of primarily woodland habit, can take advantage of a wide range of habitats. It is a rapidly maturing and fast-breeding species which can colonise new areas very quickly. As a result of its opportunistic nature and rapid dispersal, it is both abundant and widely distributed throughout Europe and Asia. Some 70–75 cm at the shoulder and weighing between 17 kg and 25 kg, roe have foxy-red summer coats, with grey face and chin, and with a pronounced black band from the angle of the mouth to the nostrils resembling a moustache. There is a white patch of hairs on the rump which can be erected and are puffed up in alarm. Neither sex has a visible tail, but in winter (when the general coat becomes duller and greyer) females grow prominent anal tufts which may be mistaken for a tail. The antlers of the male are erect, and consist each of a short main spike (typically 15–20 cm long) with at most two short branches. The typical European roe (*C. c. capreolus*) is found in almost every country of western Europe. Further east, across the Ural mountains, it is replaced by the larger Siberian roe (*C. c. pygargus*), standing some 90 cm — a good 15 cm taller than the western race — and weighing almost twice as much; while in China and through the greater part of the Korean peninsula occurs a third subspecies, the Chinese roe deer (*C. c. bedfordi*).

The most abundant and widespread of the New World deer are the two species of the genus *Odocoileus*: the white-tailed deer (*O. virginianus*) and the black-tailed deer or mule deer (*O. hemionus*). Mule deer are found over a vast expanse of western North America, although they appear to be confined to the western states, extending northwards from central Mexico to southern Alaska, and eastwards as far as Minnesota. They are found in a variety of habitats, from high mountains to plains and deserts. These are medium-sized deer of about 100 cm at the shoulder and weighing around 100–120 kg. The coat — rusty-red in summer, brownish-grey in winter, with a whitish face and throat and a black bar around the chin — is rather reminiscent of that of the European roe. The underparts and insides of the legs are white, as is the rump patch. The long tail is also white, with a black tip (extending up the outer surface in the black-tailed subspecies, *O. h. columbianus* and *O. h. sitkensis*). The antlers of the males are most distinctive, with branches not as tines from a main beam, but as symmetrical forks at each division (Figure 2.3). The

Figure 2.3 Antlers of mule deer and black-tailed deer show dichotomous pattern of branching

white-tailed deer (*O. virginianus*), with 38 subspecies, is more widely distributed in North and Central America: it is resident in perhaps every state of the US except Alaska, California, Nevada and possibly Utah, and extends north into southern Canada. The range also reaches southwards through Mexico and Central America, in the northern part of the South American continent extending to Peru and Brazil. The species has also been introduced into Scandinavia and New Zealand. In North America it is primarily a woodland species, while in the south it is restricted to river valleys near water in the dry season, withdrawing to higher altitudes during the rains. The coat is similar to that of the mule deer, but there is more white about the throat and inside the ears, and the underside of the tail is pure white. Antlers branch from the main beam, and not dichotomously like those of mule deer or black-tailed deer.

Although a number of species of both New World and Eurasian deer have been introduced into the southern hemisphere by Man, the only deer (with the exception of the Bawean deer) occurring naturally south of the equator are found in South America. There are in fact some ten species native to this area. In forested areas of Central and South America are found a number of species of brocket deer, *Mazama* spp. Characteristic generally of dense forest thickets, they are small to medium-sized deer; males bear antlers, but these are simple spikes of 10–14 cm. The red brocket (*M. americana*) (17–20 kg) occurs from Mexico through Central America to the south; one subspecies or another (there are 14) is found within every country south to Argentina, with the exception of Chile. Its cousin, the brown brocket (*M. gouazoubira*), represented in ten subspecies, has a broadly similar distribution, but tends to occupy more open forest than the red brocket. Two further species have a more restricted distribution. The little red brocket (*M. rufina*) is, as its name suggsts, a slightly smaller deer from the forest thickets of north Venezuela, Ecuador and southeast Brazil, while the dwarf brocket (*M. chunyi*), first described in 1959, is restricted to north Bolivia and Peru. Little is known of the behaviour and ecology of these species, although studies by Bodmer (unpublished) in the Amazon suggest that both *M. americana* and *M. rufina* feed almost exclusively on fruit, with 80–90 per cent of their diet consisting of fruit pulps and fruit seeds.

In the deep forests of the lower Andes are two species of pudu, the smallest of the South American deer. Standing 35–40 cm at the shoulder and weighing 9–11 kg, they are in fact about the same size as the Chinese water deer *Hydropotes*. Males carry antlers, but these are simple spikes like those of the brocket deer. The southern species (*Pudu puda*) is found in Chile and Argentina, being replaced in Ecuador, Colombia and the extreme north of Peru by the slightly larger northern pudu (*P. mephistopheles*). Little is known of either species, for they are extremely shy animals of deep forest.

There is one further group of four deer species occurring only in southern America: the marsh deer, the pampas deer, and the Chilean and Peruvian huemuls. All four are endangered or severely threatened. The marsh deer (*Blastocerus dichotomus*) is restricted to southeastern Peru, Bolivia, Brazil south of Amazonia rainforest, Paraguay and northern Argentina. As its name suggests, it is an open-country species, occurring in seasonally flooded grasslands along major watercourses. It is the largest of the South American deer, standing about the height of a small red deer. In many ways similar to the Asian barasingha, it has a shaggy chestnut coat in summer (which moults to a browner shade in winter), a bushy tail, and large ears filled internally with white woolly hair. Like many other odocoilines (roe deer, white-tailed deer, mule deer), it has a black band on the muzzle and upper lip. Relatively little is known of the behaviour and ecology of this species; numbers are, however, low and thought to be declining (Cowan and Holloway, 1978).

The related pampas deer (*Ozotocerus bezoarticus*), another open-country species, this time characteristic of seasonal grasslands, is a somewhat smaller deer (about 70 cm at the shoulder: about the size of a European roe deer). It is light in colour, being yellowish-brown on the upperparts with the insides of the ears and the underparts white; there is a line of dark brown hairs on the upper surface of the tail. Males have short, straight antlers of up to three branches, very like those of roe deer. Three subspecies of pampas deer are recognised. The typical form (*O. b. bezoarticus*) occurs on open plains through central and northern Brazil, being replaced in northern Argentina, Paraguay and western Brazil (to the River Uruguay) by another race, *O. b. leucogaster*. In Argentina and Uruguay it is represented by a third form, *O. b. celer*. This last race occurs in only four localities in the Buenos Aires and San Luis provinces of northern Argentina, with total numbers estimated in 1978 at 200 (Jackson, 1978); in Uruguay, numbers are estimated (Jackson *et al.*, 1980) at fewer than 1,000 animals. The decline in numbers is attributed to overhunting in the past, coupled with present-day erosion of habitat and competition with domestic livestock (Jackson, 1978).

The two American huemuls are both open-country or shrubland species of the High Andes. The Chilean huemul (*Hippocamelus bisulcus*) is restricted to Chile and western Argentina, while the Peruvian huemul

(*H. antisensis*) is found in Ecuador, Bolivia, Peru and the extreme north of Argentina. The two species are separated by a wide gap represented by that part of Chile north of Colchagua province, and by the provinces of Mendoza and San Juan in Argentina (Whitehead, 1972); intergradation is therefore impossible. Huemuls are solidly-built animals of about 90 cm at the shoulder. They have large, mule-like ears; males bear simple forked antlers with a single branch. In colour they appear dark brown, but, according to Whitehead, at close quarters this resolves as a mottled black and yellow, with a white tip to the chin and the lower part of the tail. Studies of Chilean huemul (Povilitis, 1978) suggest that typically they occur between 1,450 m and 1,700 m on northern and western slopes, preferring shrubland (65 per cent of observations) to forest (20 per cent) and low vegetation (15 per cent); Whitehead (1972) notes that the Peruvian huemul enter woods even less frequently than the Chilean animals.

Two further species are included within the Odocoilinae, and both occur in both the Old and the New Worlds. The heaviest of living deer, the moose or, in Europe, the elk (*Alces alces*), is a huge creature, standing at from 180 cm to 200 cm at the shoulder and weighing from 400 kg to 800 kg (males are typically some 25 per cent larger than females). A ponderous animal of northern forests, the moose is found throughout northern Europe, from Scandinavia as far eastwards as eastern Siberia, and on into northern America, where it occurs throughout Canada, Alaska and the northeastern United States. Moose are distinguished by their great size and extremely long legs (whose sinews and tendons click characteristically as they move). The shoulders appear hunched, the broad muzzle is pendulous and overhangs the lower jaw, and both sexes have an obvious dewlap of loose skin and hair hanging from the throat. The males carry huge palmate antlers, which in the European race may have a spread of 120 cm. Curious and rather solitary animals, they feed extensively on aquatic and wetland vegetation of northern forests as well as browse, and have somewhat splayed hooves to assist them in walking over swampy ground.

Finally, in our whistle-stop tour of the world's deer we must introduce the only living species in which antlers are borne by both the sexes. Another species which occurs throughout the north of Europe and across into North America, the reindeer (European) or caribou (American), *Rangifer tarandus*, is a creature of the extreme north, from Scandinavia, to Spitsbergen, European Russia and across into Alaska, Canada and Greenland. Reindeer have also been introduced onto a number of islands (for example, to South Georgia in the Antarctic) to provide a living larder of fresh meat for passing whaling vessels. This is primarily a woodland species and many populations are restricted to such habitat; but, in those geographic areas of their range where woodland grades in the north to taiga and open tundra, some populations are migratory, undertaking regular annual migrations up to the open tundra for the brief arctic summer before returning south to the forest zone to overwinter. Reindeer are large deer, 100–130 cm at the shoulder (male) and weighing 100–270 kg. Females are slightly smaller than males (shoulder height 95–115 cm), and their antlers tend to be smaller with fewer branches. The coat is brown in summer, greyer in winter, with white on the rump and tail and above each hoof. Mature males develop a heavy white mane during the breeding season (or rut). Some nine subspecies are recognised; many of the European herds are semi-domesticated, herded for meat and skins.

(References: While much is known of the biology of Odocoilines such as the moose, reindeer, roe and mule and white-tailed deer, disappointingly little is known of many of the South American species. Useful summaries of the biology of roe deer will be found in Anderson, 1961; Prior, 1968; Strandgaard, 1972; Delap, 1978. Mule deer: Taylor, 1956; Walmo, 1981. White-tailed deer: Taylor, 1956; Hirth, 1977; Halls, 1984. Moose: Petersen, 1955; Franzmann, 1978. Reindeer: Bergerud, 1978; Leader-Williams, 1988. Brief accounts of the other species are probably best found in Whitehead (1972) and IUCN (1978). A review of population ecology of brocket deer in Suriname is presented by Branan and Marchinton (1985).)

Although this volume is to be concerned primarily with living deer, any natural history of the group would surely be incomplete without some reference to the more memorable of the extinct species, and we should perhaps not leave this section without reference to the largest deer that has ever lived, the giant Irish 'elk' (*Megaloceros giganteus*) (Figure 2.4).

Figure 2.4 The extinct giant deer, or Irish elk, whose antlers were so large they probably weighed more than the rest of the entire skeleton

Fossil antlers of this giant deer have long been known in Ireland, where they occur in lake sediments underneath peat deposits. Before attracting the attention of scientists, they had been used as gateposts, and even as a temporary bridge to span a rivulet in County Tyrone (Gould, 1973; 1977).

Ireland's exclusive claim vanished in 1746 (although the name stuck) when a skull and antlers were unearthed in Yorkshire, England. The first Continental discovery followed in 1781, from Germany, while the first complete skeleton (still standing in the museum of Edinburgh University) was exhumed from the Isle of Man in the 1820s (Gould, 1973). It appears that these giant deer evolved during the glacial period of the last few million years, ranging as far east as Siberia and as far to the south as northern Africa. They may have survived until historic times in Continental Europe, but became extinct in Ireland itself about 11,000 years ago as a result of climatic change (Gould, 1973; 1977).

REPRISE: WHAT MAKES A DEER A DEER?

The deer family clearly consists of a wide diversity of species adapted variously to widely differing environments. All are, however, herbivores and, although it is developed to different degrees in different species, all possess a rumen for digestion of plant cellulose. All share with other large herbivores adaptations for avoiding predation: acute senses, and modifications in the skeletal and muscle structure of the limbs to enable them to run rapidly (pages 4–5).

In most species males carry antlers, although in the more primitive forms (musk deer, Chinese water deer) these are absent, and in one group (the reindeer) they are borne by both sexes. In those groups which lack antlers, the upper canines are instead well developed as sharp tusks. Such tusks are also retained by the muntjac, although these are not so well developed as those of the musk deer or the water deer. Upper canines are also present in some other deer (Père David's deer, red deer, wapiti and others of the genus *Cervus*), although in these more advanced species they are not developed as tusks. In all remaining species they are normally absent altogether.

The canines of the lower jaw are invariably present in all deer, but have been modified in such a way as to become effectively extra lateral 'incisors', and in fact the dentition of the deer is very simply divided into biting incisiform teeth of this sort for plucking vegetation and grinding cheek teeth (molars and premolars) for mastication of the food. Incisors are present only in the lower jaw and the animal bites against a special horny pad on the upper jaw. The general dentition can be represented thus (described as for one side only):

Incisors	Canines	Premolars	Molars	Total
$\frac{0}{3}$	$\frac{0}{1}$ or $\frac{(1)}{(1)}$	$\frac{3}{3}$	$\frac{3}{3}$	$\frac{6}{10}$ or $\frac{7}{10}$

Not all the deer develop their full mouth of permanent teeth in the same period of time. At birth, all have six incisors and two lower canines, but the replacement of these milk teeth and development of the full adult dentition depends very much on the species. The red deer for example requires some 30 months to develop a full mouth, while the European roe takes only 13 months (Whitehead, 1972).

All deer possess various scent glands, used for recognition, for territorial advertisement, and for sexual advertisment: thus they all have glands on the face in front of the eye, and most also have interdigital or intertarsal glands opening into the cleft of the hoof. Many species also have a gland situated on the hindleg, with the position marked by tufts of hair longer than those of most of the surrounding area. Finally, males of most species possess a scent gland associated with the prepuce, or penis sheath, which becomes active during the breeding season (it is this gland which is so developed in the musk deer for the production of musk); and, in some members of the Odocoilinae at least (roe deer, white-tailed deer), males appear also to have scent glands on the forehead, used for rubbing on trees and other vegetation. Both sexes possess teats, and here again musk deer differ from other deer, as already mentioned, in that they have only a single pair of teats while all other groups have four.

We have noted that among the deer there tends to be a size difference between the sexes. In musk deer the females tend to be slightly heavier than the males, but in all other species it is the males that are larger. The degree of this dimorphism is variable (in muntjac or roe deer, for example, the sexes are of more or less equal size and the difference in weight is only of the order of 20 per cent, but in a species such as the red deer the difference may be as high as 70 per cent: Clutton-Brock *et al.*, 1982), but it appears to be related to the type of mating system adopted, with the degree of dimorphism related to the degree of polygyny. In those species where a single male may monopolise a harem of many females, competition between males for ownership of females (and chance to breed at all) becomes intense. In such competition, larger males will be at an advantage; selection will favour larger individuals, which will ultimately lead to an increase in male size within the species as a whole. It would appear that the size differential between the males and females has thus arisen in response to the degree of sexual competition experienced within the particular mating system adopted.

These differences between the sexes in size, and in possession of antlers, is in fact reflected in the different common names given to males and females: thus throughout the genus *Cervus*, and in many other species, males are referred to as 'stags', females are called 'hinds', and young are referred to as 'calves'. Unfortunately, however, the system is not universal: and in other species — fallow deer, white-tailed deer, muntjac, roe — males are termed 'bucks', females 'does', and the young 'fawns' (or even, in the case of roe, 'kids'). Just to make things worse, in moose and reindeer and wapiti, males are bulls, females cows, and the young once again are calves.

Although the system is no longer consistent — and which species fits which classification is somewhat arbitrary — in its original conception the overall scheme derives from the ancient traditions of hunting, when 'Royal' beasts of the Forest (red deer and their relatives: animals which could by ancient law be hunted only by the monarch) were referred to as stags and hinds, to distinguish them from common beasts of the Chase (fallow and roe) which could rightfully be pursued by lesser nobles (see pages 156–7). The hunting tradition has in fact also provided a whole series of titles to be applied to male deer of different ages within each species. Thus, within fallow deer, males in their first, second, third, fourth and fifth years are, respectively, fawns, prickets, sorels, sores, and bare bucks, while adult males of six or more years old, in the glory of full antler growth, are deemed 'bucks' or 'great bucks'. These finer distinctions are not preserved within the pages of this book, for they serve merely to confuse those unfamiliar with the terms. Males, females and young will, however, be referred to as stags and hinds, bucks and does, calves and fawns, in accordance with normal practice for the species concerned.

EVOLUTIONARY RELATIONSHIPS AMONG THE DEER

Evolutionary relationships between the major groups and subfamilies of living deer seem fairly clear (Figure 2.5). The musk deer, with their more simple stomach structure and lacking antlers, but with enlarged canines, show striking similarities to chevrotains (page 2), and of all the living deer seem to show the closest relationship to this group and their traguloid ancestors. Such relationship suggests that the musk deer may have developed as a relatively early side shoot of the line leading to the more advanced groups of deer. The single pair of teats and the presence of a gall-bladder (more a bovid than a cervid character) also suggest that this is a group arising fairly early from the main line of cervid evolution. Lack of antlers and the possession of tusk-like upper canines also seem to cast the Chinese water deer as a relatively primitive group, although other characters are more certainly cervid, implying that while these deer may also represent an offshoot from the main line of cervid development they trace their origin from a somewhat later development of that evolutionary line. Muntjacs — again an Asiatic group — with a more advanced stomach structure and with simple antlers, have clearly progressed further along the line to the advanced deer, but the relatively elongated pedicles, the simplicity of the antlers and the retention of an upper canine tusk, albeit shortened, still cast these as a relatively early group. Finally, the Eurasian Cervinae and the, predominantly New World, Odocoilinae obviously represent two highly successful and relatively independent radiations among the more advanced deer.

Evolution throughout appears to have been accompanied by general increase in body size, in size and complexity of antlers, and in general a tendency towards a more gregarious habit.

Although the general relationships between these major subfamilies seem clear enough, details of relationships *within* the larger groups (Cervinae and Odocoilinae) are obscure. The small size and simple antler form of brockets and pudu may suggest that these are fairly primitive members of the Odocoilinae, yet could equally represent secondary adaptation to the dense forest environments in which they are found, so that the relationship of these two genera to others in the subfamily is certainly ambiguous. Both the moose and the reindeer are now so highly specialised that it is equally hard to trace their origins. In both, structure and physiology are so dominated by features which are clearly specialist

Figure 2.5 Evolutionary relationships between the living families of deer

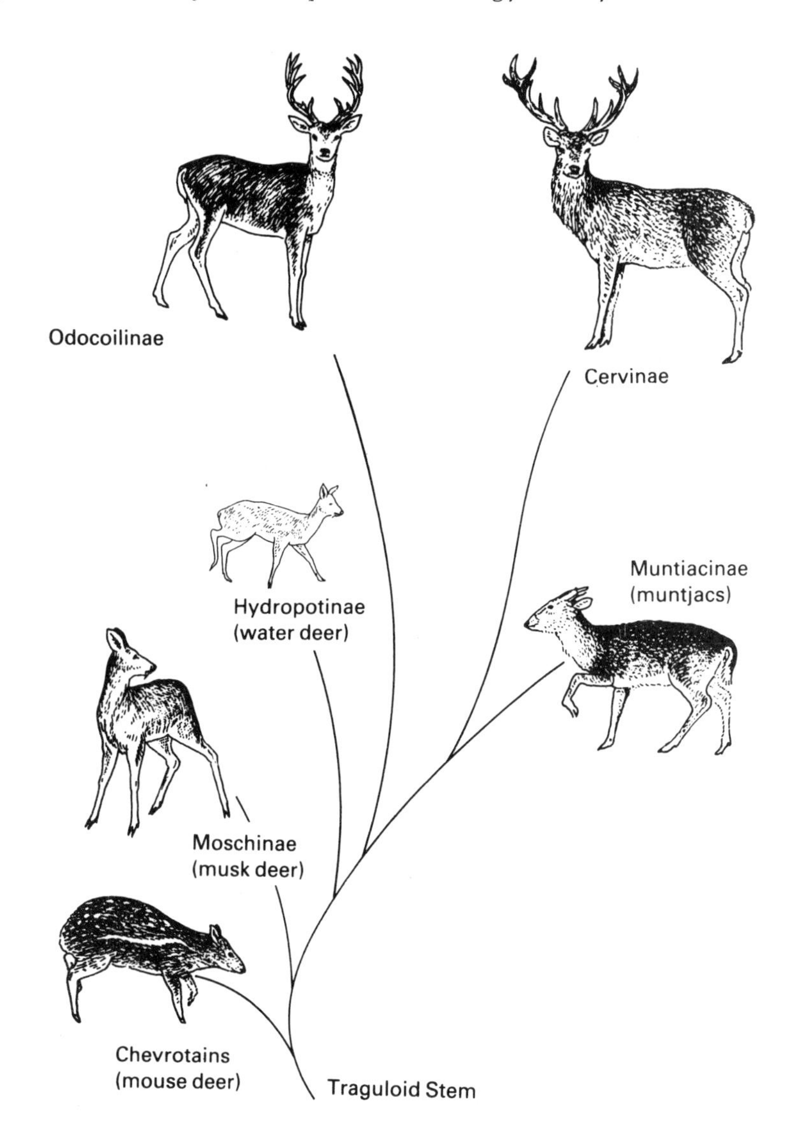

adaptations to their current habitat and life style that any clues to their original ancestry are well masked.

Among the other Odocoilinae, however, we may with the eye of faith perceive a hint of pattern. All are clearly closely related. Hybridisation has been recorded in captivity and in the wild between mule deer (*Odocoileus hemionus*) and white-tailed deer (*O. virginianus*); indeed, Stubblefield *et al.* in 1986 considered that, in some counties of Texas, occurrence of natural hybrids could be as high as 13.8 per cent. Other, morphological and structural, characteristics also suggest close affinities: both *Odocoileus* species, together with marsh deer (*Blastocerus dichotomus*) and pampas deer (*Ozotocerus bezoarticus*), share with the European roe deer a light patch on the lips and throat with a clear 'moustache' or bar of dark brown around the muzzle and chin (Plates 13, 14)

Various features suggest that the roe deer of Europe is an ancient species. It is small and in general less social in nature than many of the other species, it has quite simple antlers and, most strikingly, has a simpler stomach structure than other Odocoilines. It seems possible that some similar form, again originating in Europe, spread across to North America at some time in the Pliocene, and from that stem developed the more advanced white-tailed and mule deer and possibly also the marsh deer, pampas deer and huemuls, these latter species representing radiations from the basal stock in specific adaptation to the various different habitats — marshes and floodplains, open grasslands, high montane environments — to which each has become specialised.

If anything, relationships between the various species of Eurasia are even more confused. Studies of chromosome number and various other features suggest that the curious Père David's deer (currently set aside in its own genus *Elaphurus*) is at the least very closely related to the genus *Cervus*, to the extent that some authorities suggest that it should be included within that latter genus. We should perhaps note at this point that hybridisation between Père David's deer and European red deer has been reported on several occasions (Harrington, 1985) and indeed is now being used as a regular feature in breeding programmes designed to 'improve' farmed red deer stock (e.g. Tudge, 1987). Such relationship leaves us essentially with three genera to consider: *Dama, Axis* and *Cervus*. All are clearly themselves closely related. Hybrids have been reported between fallow deer and hog deer (Van Ee in Gray, 1971), and at various times fallow deer and the *Axis* group have been placed in the genus *Cervus* with all the other Cervinae. Details of origins and relationships, both between these genera and between congeneric species, are, however, obscure, and nowhere more confused than in the genus *Cervus* itself. While certain species such as the brow-antlered deer (*Cervus eldi*) or Thorold's deer (*C. albirostris*) are quite distinct and clearly represent local and highly specialised variations on the theme (*C. eldi* for example, or *C. duvauceli*, both local specialisations for swamp-dwelling habitat), others are basically 'red deer' with little claim to special status as distinct species beyond geographic isolation. Almost all hybridise freely, with the production of fertile offspring, where artificially or naturally (at the borders of abutting ranges) they are brought into contact. With such lack of reproductive incompatibility, it is probable that many species currently recognised within the genus are in practice only of subspecific status, if indeed that.

Hybridisation occurs readily between red deer and wapiti and red deer and sika, although gross differences in size between the wapiti and sika prevent such hybridisation without artificial insemination. Such observations led Harrington in 1982 to suggest that these three species at least represented a ring species: a single species with variation around the globe whose adjacent neighbours could interbreed freely, but at whose extremes the forms had diverged so far that, even when they do co-occur (as wapiti and sika do in eastern Asia), reproductive isolation was established. Such a concept would establish a single species, *Cervus elaphus*, with wapiti, sika and perhaps others, such as sambar and rusa, as mere subspecies.

(A parallel situation exists in Eurasian great tits, *Parus major*, where a series of populations extends from the east coast of Soviet Russia, through Russian itself to western Europe, thence south of the Caspian Sea, through to India and on northwards through China and Korea back to the eastern coast of Russia. While the various populations around this circuit differ from one another, they intergrade, and it is clear that adjacent, neighbouring populations regularly hybridise. However, by the time the circle is complete, where the two *ends* of this continuous ring meet at the Amur River in USSR they have diverged sufficiently that they themselves cannot interbreed, despite the fact that they intergrade with each other through neighbouring populations right around the ring and so strictly might be considered one biological species.)

The similarity of many of the mainland Asiatic sika deer to contemporary hybrids between red deer, *Cervus elaphus,* and Japanese sika, *C. nippon nippon* (now regularly recorded in the UK and Germany, where introduced sika deer have invaded native red deer range, and in New Zealand, where both species have been introduced), led Lowe and Gardiner (1975) to propose a slightly different scheme suggesting that, while red deer and the Japanese sika might represent true and distinct species, many of the intermediate forms currently recognised on the mainland as distinct subspecies might themselves be of past *hybrid* origin, probably a result of introductions due to Man's ancient trading interests. The rapidity with which hybridisation and substantial genetic introgression can occur between the two species, and the fact that the hybrid appears to have an adaptive advantage, at least over the Japanese form, in some areas, however, support the idea that the Japanese deer themselves are truly of little more than subspecies status (Harrington, 1985). By the same token, then, clearly the wapiti, too, must also be accorded only subspecific status (a conclusion also supported by other genetic studies of blood proteins: e.g. Dratch and Gyllensten, 1985). Harrington goes on to argue that, if this is so, even despite the major differences in form between Japanese sika deer and European red deer, there should be an even closer relationship between the less different 'species' of this genus: for example, *Cervus unicolor* is also known to hybridise successfully with red deer (Powerscourt, 1884).

Lydekker (1898) considered the sika deer to have a common origin with wapiti but to have evolved first from that lineage. The evidence from serum proteins supports this (Harrington, 1985). This same immunological evidence also supports the idea that the sika deer inhabiting those areas not within natural colonising distance of the Japanese islands are of recent man-made hybrid origin. It is possible that the rusa deer of the Indonesian archipelago have originated from sambar deer in the same way.

Gray (1971) reported hybridisation between *Axis axis* and *A. porcinus* in which the F_1 offspring were fertile; in addition, as we have already noted, *A. porcinus* was reported by Van Ee (cited by Gray, 1971) to have hybridised with *Dama dama*, the fallow deer. Such intergeneric and interspecific hybridisation suggests genetic inter-relationship within this group, too, similar to those within *Cervus*.

3 The ecology of resource use

PATTERNS OF RESOURCE USE

While the deer family indubitably evolved primarily as a group of species of woodland or woodland edge, in practice they show a wide variety of adaptation to differing habitats. While many modern species remain woodland animals, others have become adapted to other kinds of ecosystem, colonising such specialised environments as swamplands or marshy ground (as have the moose, barasingha and thamin) or, like the pampas deer of South America or hog deer of Asia, secondarily adapting to life in open grassland. Others again, as for example the roe, have shown themselves highly adaptable and able to colonise a whole variety of different environments from dense woodland to open range. Further, while most species remain specialists, the majority faithful to their woodland origins, even here there is tremendous variation: there are, after all, so many different types of forest within their geographic range and so many different life styles possible even within one forest type.

Deer are found in forests or woodlands from the tropics to the temperate zone and north into the artic taiga. They are found in evergreen forests of conifer, broadleaved deciduous forest, thickets of dense bamboo; in even and equable environments, through strongly seasonal ones — such as the seasonal rainforests of Asia or those in temperate climates in both hemispheres subject to the seasonality of summer and winter — to the extreme environments of the far north. They are found in lowland forests and up in the high montane forests of India or of Peru and Chile. Some lead secret lives in the middle of the densest forest; others prefer open-based woodland or woodland edge. Even within each forest type there is plenty of room for variation, as different species feed in different ways on different types of vegetation (lichens and mosses, grasses or broadleaved forbs, woody browse) and, depending on their size, feed at different heights above the ground. In short, even though the deer are primarily a woodland group of species, in practice they occupy a bewildering variety of different environments and must adapt to a wide range of different pressures.

Such adaptation is apparent at two main levels. First there will be general characteristics of behaviour, physiology or morphology to 'fit' the animal to the general demands of the habitat and life style it has come to adopt. Such adaptations are, however, but loosely tuned, for they accumulate through evolutionary time and, in effect, adapt the animal to an 'average' set of conditions that the population might expect to encounter over the whole of its range and over a number of years. Each individual animal, however, has to cope with the conditions it experiences at any one instant; thus, while evolution has equipped it with a general adaptation to the sorts of conditions it is likely to experience overall, it must make finer adjustment within that, by behavioural adaptation, to cope with the actual conditions experienced. All animals must seek from their environment all the necessities of life — food, water, shelter from the elements, shelter from predators; and patterns of resource use observed can be explained in relation to the satisfaction of these requirements. Evolutionary adaptations for the species as a whole enable it to cope with gross features of the environment: gross climate, general habitat type, general diet. Behavioural adaptations (use perhaps of slightly different sub-habitats for food, shelter, and avoidance of predators in relation to the different qualities of each, or changes in patterns of habitat use and diet through time as the resources on offer themselves vary through the course of a day or between the seasons) enable the individual animal to satisfy its actual day-to-day requirements in relation to the particular characteristics of the specific area in which it finds itself.

In this chapter we shall consider the adaptations shown among the deer to their particular environments in relation to their pattern of use of habitat, food resources and time.

HABITAT PREFERENCES

While most of the world's deer are, as we have noted, essentially species of forest, woodland or woodland edge, some have largely left behind their woodland origins and moved out to colonise entirely different types of environment. We have already described how barasingha and thamin of central Asia and the marsh deer of South America have specialised in swampy habitats or marshlands. The Chinese water deer favours ricefields or other open grasslands in the Far East, where the tall grasses provide for it both food and shelter; in India and central Asia, this type of habitat is taken over by the hog deer, which is characteristic of the grass jungles of the Indian subcontinent. In the New World, once again we find an 'open country' species in the pampas deer of South America. In the mountain regions of Asia and South America we find specialist species such as the various species of musk deer, and the two huemuls of the High Andes. All these species are clearly habitat specialists, adapted in various ways — in structure, physiology and behaviour — to their particular specialised environment.

Examples such as these are perhaps somewhat artificially clear-cut, since we have been specifically considering animal species which have specialised to unusual or extremely distinctive types of environment. Even among those we might consider as 'woodland' species, however, we can see clear evidence of specialisation to particular *kinds* of woodland, preference for particular woodland structure. Thus, as we have already noted, some species favour dense forests, others open woodland or woodland edge. Very few are complete opportunists, and it is usually possible to define for any species quite a narrow range of conditions in which it may be found. Perhaps the clearest illustration of this comes from situations where a number of deer species all occur together within a single geographic area. In such a case there can be no obvious separation between deciduous and coniferous forest, tropical or temperate. Yet differences do exist in the habitats favoured by the different species, and subtle differences in habitat preference do serve to separate them from each other. In European forests, for example, we may find red deer, roe, fallow and, where it has been introduced, sika deer all occurring together in one locality; in south-central Asia we may find together chital, sambar, hog deer and Indian muntjac; in the forests of Central and South America four species of brocket deer have overlapping ranges.

In each case, closer examination reveals that every species has its own distinct habitat preferences, and performs at its best in a subtly different set of conditions from those which best suit others. Brocket deer, which differ primarily in size, separate out largely in altitude, and in forest density; in European forests, red, roe and fallow use the mosaic of different sub-habitats in different ways. Fallow tend to prefer deciduous and mixed woodland blocks, lying up in denser cover, moving out to feed in open areas — clearings or nearby agricultural fields; they are perhaps more species of open woodland/woodland edge than the other two, and, as we shall see, feed preferentially as grazers. Red and roe are more characteristic of the woodland core, where in fact both prefer dense woodland or thicket for cover, resting up in dense cover, moving out to feed in slightly more open habitat, and feeding on browse materials or the grass and herbs of the woodland floor. Even here, however, there are subtle differences. Roe are small deer (about 20 kg against the red deer's 150–200 kg) and tend to be specialist feeders. Because they select highly nutritious herbs and forbs — or growing shoots of browse — and because the body size is small, they do not require a massive food intake and can satisfy their needs within a relatively small home range. The larger red deer ranges more widely and is a less selective feeder. With a greater body mass (and smaller relative ratio of surface area to volume) it does not need to select so carefully, for its cover habitats, those which offer the absolute maximum protection from wind and weather; and as a result, once again, although their use of habitat is similar, the actual preferences of the two species and pattern of habitat use expressed are subtly different (Figure 3.1).

In the same way, in the Chitawan National Park of Nepal, Mishra (1982) has studied the habitat use of chital, sambar, hog deer and muntjac. All occur within the park, which offers to the deer areas of grassland, riverine forest and sal forest. Mishra found, during three years of study, that 75 per cent of his observations of chital were in riverine forest, with 23 per cent in grasslands and only 2 per cent in sal forest; by contrast, muntjac, which made extensive use of riverine forest (78 per cent), made little use of grassland (7 per cent) but much more use of sal forest (15 per cent). Sambar showed a pattern of habitat use not unlike that of the muntjac (80 per cent riverine forest, 10 per cent sal forest, 10 per cent grassland) but clearly differ in size, and use different areas within the forest from the smaller muntjac; and the little hog deer is, as

Figure 3.1 Habitat preferences of red and roe deer in coniferous forest. The figure shows the proportion of observations of each species in different habitat types. Source: Hinge (1986)

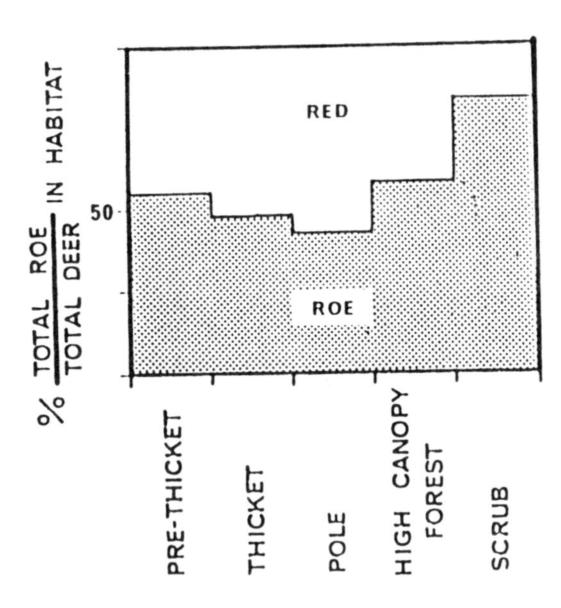

we have already noted, a grassland specialist, with 95 per cent of all observations made in open grassland. Similar separation in use of different grassland types between chital, hog deer and barasingha in another area of Nepal have been recorded by Schaaf (1978). Even within a given area, then, the different species are using the environment in different ways, specialising in slightly different habitat types.

We should note that the precise habitat occupied may of course vary from place to place as climatic conditions change, or as the community itself varies and any one species finds itself living in association with different combinations of other deer. Thus, while Mishra's observations of chital in Nepal suggest that they are a species characterised by a mixed use of riverine forest and grassland, elsewhere in their range they may be more characteristic of true forest (as in northwest India, where they appear to be true forest dwellers: Berwick, 1974) or by contrast may be associated more with open environments (as in central India, where they are found to make much more extensive use of open meadows, grazing upon the short grasses: Schaller, 1967). Sambar, too, while primarily an animal of forests and dense woodland, may be found living in large groups in more open country in Sri Lanka; and, while hog deer are more usually associated with the open grass jungles of Asia, they too may also be found — solitary and territorial — within dense forest habitat (Kurt, 1978).

Some species have developed this adaptability to a considerable degree and become true opportunists. The roe deer of Europe, while originally a woodland species, is a relatively small species which can satisfy its needs from a small home range. All it really needs is the smallest amount of cover and an area of rich feeding (it requires only a comparatively small quantity of food, but that food must be of high nutritional value). Within woodlands, therefore, it tends to be associated with gaps in the canopy, with regeneration (or Man's plantations in commercial forests) and the earlier stages of ecological succession. Even a ditch or a hedgerow, however, may provide sufficient cover, and Man's agricultural crops, by definition, tend to be fast-growing and of high nutritional quality. Thus, even where woodlands may be few and far between, roe can often survive and prosper. In the cereal provinces of central Europe, vast expanses with barely even a hedgerow, roe can still be found, indeed found at extremely high densities (e.g. 12 animals per km^2 in agricultural areas of southern Czechoslovakia: Zejda and Homolka, 1980).

Such opportunistic species — as roe or hog deer — or those already adapted to grassland habitats, like the Chinese water deer, are clearly better able than extreme woodland specialists to withstand and surmount man-imposed changes to their environment; indeed, they can in certain cases even take advantage of such alteration, as we see here, exploiting to the full the new habitats created by Man's activities (see Chapter 8).

PATTERNS OF HABITAT USE

While a species as a whole, or a particular geographical population, may be characteristic of some general environmental type, any environment is in fact composed, at finer level of resolution, of a complex mosaic of different habitats and sub-habitats. Individual animals will thus use these different components in different ways, seeking different resources from different parts of the whole mosaic; this pattern of habitat use will itself change through time as the requirements of the animal or the resources offered by different vegetation types change by day and night, or through the different seasons of the year.

We have already seen, for example, that chital in Nepal do not use just one vegetational community, but spread their time between grasslands, riverine forest and sal forest: overall, spending 23 per cent of their time in grasslands, 75 per cent of their time in riverine forest and only 2 per cent in sal forest. Mishra (1982) noted in addition that the pattern of use changed with season just as we have suggested. Chital showed their least use of riverine forest and greatest use of grasslands just before and during the monsoon period, when burning of dry grassland vegetation and the regrowth during the rains meant that these areas offered better forage. Studies of the same species in Ruhuna National Park in Sri Lanka by Balasubramaniam *et al.* (1980) showed even more striking variation between the seasons in both habitat use and diet: in the dry season most of the deer were associated with forest/scrub habitats, and were observed to be browsing extensively on shrubs and trees, while, in the wet season, the majority of deer were found grazing in the open plains.

Seasonality in habitat use in relation to food availability and food quality is of course especially marked in areas with distinct wet and dry seasons of this sort, and the shifts are to a great extent imposed upon the deer by seasonal changes in the plant communities themselves. As the water table drops lower and lower during the dry seasons, grasses and forbs and all shallow-rooted herbaceous vegetation die off and it is only the deeper-rooted trees and shrubs that still provide acceptable fodder. Such seasonal change in habitat use among tropical and subtropical species, while it indubitably does reflect changing availability of foodstuffs, may, however, also be influenced by another critical factor: availability of water. Schaaf (1978) showed clearly that seasonal range (and resultant habitat use) by barasingha in Nepal was strongly affected by the need for access to water during the dry season.

Where seasonality reflects a change in climate not between wet and dry, but in temperature, yet another factor may combine with changing food availability to influence habitat selection: shelter. The pattern of habitat use by fallow deer in the New Forest of southern England, for example, is strongly seasonal and can also be shown to be affected primarily by seasonal availability of different forages. Thus fallow deer are preferential grazers, and grasses constitute more than 60 per cent of their diet for much of the year. The short growing season of temperate zones, however, limits the period of grass growth to a few months of the summer. During autumn and winter, the deer must exploit other foodstuffs: the abundant but ephemeral fruit crop of the forest trees (acorns and beech mast) and browse materials (leaves and twigs.) Habitat use reflects this (Table 3.1). Overall, the fallow make greater use of

Table 3.1 Habitat use by New Forest fallow deer, expressed as percentages of observations recorded in each habitat (based on data from Parfitt, in Putman 1986)

	Woodland rides and glades	Open grasslands	Heathland and bog	Deciduous woodland	Coniferous woodland
Spring	22	2	0	50	26
Summer	42	8	0	15	35
Autumn	27	2	0	43	28
Winter	37	14	0	22	27

deciduous woodland, woodland glades and areas of grassland than of coniferous woodland or heathland, but there is clearly marked seasonal change in the pattern of habitat use expressed through the year. Deciduous woodland is actively selected in early spring (February–April) and autumn (August–October); and woodland use remains high throughout the winter in those years when the canopy trees produce a good crop of mast, as the animals remain within the woodland to exploit this abundant supply of food. In years when such mast is less abundant, use of woodland declines more rapidly during winter, and the deer make increasing use of more open habitats, grazing in clearings and glades or feeding out onto adjacent heathlands or agricultural fields; these more open habitats are again used heavily in midsummer (June/July).

While much of this change in habitat use may be explained by foraging behaviour, as in our earlier examples, in this case there is another factor contributing to the observed seasonal change in habitat selection. Fallow deer in temperate environments are exposed in certain seasons to inclement weather conditions; minimising energy loss through exposure is just as realistic an option as maximising energy gain through efficient foraging. To some extent habitat use in winter is influenced therefore by selection of cover communities; and of course the relative importance of this factor, and the degree of influence it imposes upon habitat use, increases as the severity of the environment increases (e.g. Jenkins and Wright, 1988).

At its most extreme, seasonal change in 'habitat selection' might be translated into an actual migration between two discrete geographic ranges. Barasingha in central India move between distinct dry- and wet-season ranges (Martin, 1978); northern reindeer frequently undertake lengthy migrations between north-temperate forest (their winter range) and the open meadows of the arctic tundra (their summer range). Once again, these shifts in 'habitat' can be related both to changing availability of food and to the restrictions of other considerations — availability of water in the dry season for barasingha, and increased importance of shelter in the winter for the reindeer.

Cover communities not only provide shelter from the elements for temperate or north-temperate species, they may also provide, for all species, some measure of protection from predators; and this is of course something as relevant in the tropics as in the high tundra. While predation is an ever-present threat, its importance may also vary seasonally, with risks greater in some seasons than in others. Predators may themselves be migratory, and present within the prey range only at some times of year. Predation pressure may be more intense when predators are feeding growing cubs, or the prey themselves are perhaps more vulnerable around the time when they themselves give birth or have young calves at foot. This, too, may therefore result in seasonal change in habitat use. In addition, both cover from predators and shelter in poor weather may effect temporal changes in habitat use over a shorter time-span, as different habitats may be selected for feeding or for resting and rumination.

Sika deer were introduced into the New Forest in 1908; like fallow deer, they show a distinct pattern of use of the vegetational mosaic offered by the environment (Table 3.2). In autumn and winter, like the fallow, sika favour the mature oak and beech woodland areas, where they

Table 3.2 Seasonal change in habitat use of New Forest sika deer, expressed as percentages of observations recorded in each habitat (data from Mann, 1983)

	Deciduous woodland	Heathland	Mature coniferous woodlands	Thicket and polestage conifers	New plantations	Rides and glades
Spring	52	0	0	17	10	21
Summer	34	0	0	20	19	27
Autumn	41	0	0	17	18	24
Winter	62	0	0	14	5	19

collect to feed upon the abundant mast. In late autumn and winter, as these food supplies become depleted and the weather becomes more severe (increasing the importance of shelter as a consideration in the animal's habitat selection), the deer increase their use of coniferous woodland. During the summer, use of deciduous woodland declines and the majority of the sika are to be seen in young conifer plantations and thicket areas. Once again, the change in habitat use can be explained in terms of changes in food availability and changes in the need for shelter. It is, however, not only over the course of the seasons of the year that the pattern of habitat use shows a change. There is also marked change in the habitats occupied at different times of day (reflecting the changing needs of the animal for food or cover over this shorter time-scale). In fact, Table 3.2 masks the full extent of the seasonal change in habitat use by New Forest sika, because the general change recorded there is also accompanied by a change in the times of day during which particular habitats may be used. If we look at the habitat use during hours of daylight and hours of darkness (Figure 3.2), not only do we see a marked change in the use of different habitats by day and by night, but we also see a far more striking contrast in the habitats used at different seasons. For, while Table 3.2 suggests in fact only slight change in the use of deciduous woodlands through the year, in fact the deer are found in such habitats throughout the 24-hour period in autumn and winter, but use them in summer only at night. (This change of use in itself reflects a change in what the habitat is being used for: in autumn and winter the deer are selecting the woodlands because they offer a rich food supply in the fallen acorns and beech mast, while in summer they are feeding during the day largely in coniferous plantations and using the forest's oak woods or older conifer blocks for lying up at night.)

Clearly, deer are highly adaptable and flexible, and can adjust their patterns of behaviour to show an appropriate use of the resources offered by a series of habitats, so that at any time the pattern of habitat use expressed allows them to satisfy from their environment their needs for food, water and shelter to the maximum degree possible in their given circumstances. Habitat use in a given area changes with time of day or

Figure 3.2 Temporal change in habitat use: differences in pattern of habitat use of New Forest sika during the hours of daylight and darkness (based on data from Mann, 1983)

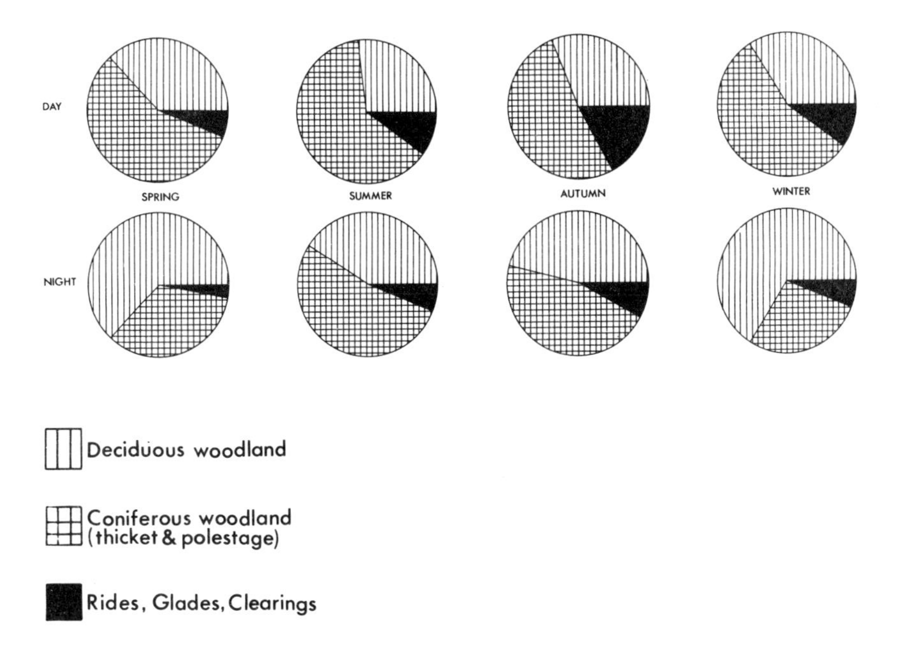

season as the needs change and the resources available to satisfy those needs also alter. Habitat use by individuals of the same species in contrasting environments will change, as, once again, needs differ from place to place and the resources available may differ. Thus habitat use of, for example, chital deer in Chitawan National Park in Nepal, reported by Mishra (1982), differs noticeably from that reported for chital in northwest India (Berwick, 1974) or in Sri Lanka (Eisenberg and Lockhart, 1972; Balasubramaniam *et al.*, 1980). In every case, however, the animal is able to adapt to satisfy its requirements from the resources on offer; and it is these needs to obtain food, water, shelter and protection that will dictate the pattern of use of resources that we observe.

We can perhaps best illustrate this by returning to Chris Mann's studies of sika deer in the south of England, for in this area populations of sika not only occur in the relatively varied vegetational mosaic of the New Forest, but are also found in the very different environment of commercial coniferous forests in Dorset. The habitats available in Dorset are totally different (and the variety on offer more restricted in the coniferous forest). Here, the pattern of habitat use observed is as shown in Table 3.3. Although the vegetation types used at different times of year differ tremendously from those used at the same times by New

Table 3.3 Seasonal pattern of habitat use by sika deer in Purbeck Forest, Dorset, England, expressed as percentages of observations recorded in each habitat (data from Mann, 1983)

	Deciduous woodland	Heathland and saltmarsh	Mature coniferous woodlands	Thicket and polestage conifers	Rides and glades	Open fields
Spring	3	27	0	31	3	36
Summer	0	12	0	57	4	27
Autumn	4	18	0	38	3	37
Winter	4	17	0	42	4	33

Forest sika (Table 3.2), the similarity in character of vegetation types favoured at different times in the two areas makes it clear that the actual vegetation itself may in practice matter very little. It is the degree to which each satisfies the animal's needs for food and shelter, and the way in which those needs themselves change through the year, that actually determine the patterns of habitat use we observe.

FOOD AND FEEDING

While all animals must seek from their environment not only food and water, but shelter and cover from predators, it is clear that one of the most significant factors affecting the patterns of habitat use expressed is the quality and availability of forage.

As we have already noted, deer are all ruminant herbivores, adapted for feeding on plant materials by various structural modifications of the gut and most notably by the possession of a complex multichambered fore-stomach for the fermentation of cellulose. Plant material, however, comes in a variety of shapes and forms — herbaceous materials, broadleaved herbs (forbs) and grasses, leaves of trees and shrubs, fruit, flowers and twigs — and different deer species have come to specialise in rather different types of diet. How have they become adapted to deal with the rather different requirements of the different food specialisms, and what factors determine which particular foodstuffs they select?

The first thing to remind ourselves is that not all deer have developed to an equal extent the full four-chambered complex of rumen-reticulum, omasum and abomasum. The more evolutionarily primitive subfamilies such as the musk deer, the water deer and the muntjacs lack the full complexity of development achieved among the Cervinae and Odocoilinae. The stomach is four-chambered as in more advanced species, but rumen pillars are poorly developed (and thus may do little to prolong the retention of fibre) and the relative size of the omasum is extremely small. It is within the rumeno-reticular complex that the active digestion of cellulose occurs; animals such as these more primitive deer, with relatively restricted development of the rumen and reticulum, are thus unable to digest these structural carbohydrates from plant material particularly efficiently. By definition, therefore, they cannot base their nutritional 'strategy' upon efficient fermentation of plant fibre; instead they must select foods low in fibre, but high in soluble carbohydrates, proteins and fats, foods which are easily digestible and whose nutrient value is concentrated in cell contents rather than within the cellulose of the cell wall. The relatively small size of these deer, which in turn must limit the potential capacity of the stomach and intestinal tract, and the relatively short time for which food is retained in the simple gut, must also limit their capacity to digest cellulose and restrict them to a nutritionally more concentrated diet. Thus we find the musk deer, muntjacs and water deer highly selective feeders, plucking small morsels of nutritious materials. The musk deer and muntjacs are forest species, feeding as selective browsers, plucking young leaves and shoots from forest trees or shrubs, and feeding in season on fruits (Figures 3.3, 3.4). Chinese water deer are of course animals of more open country, but they, too, are selective feeders, taking herbs and forbs, and young sweet grasses rather than the coarser and more fibrous vegetation of mature grasses.

Among the Cervinae and Odocoilinae, the full complexity of ruminal development is more completely realised (although even here there are differences in degree: the rather more primitive roe deer, for example, again has a less well-developed rumeno-reticular complex than some of the other species and is thus perforce something of a concentrate-selector). The deer, however, occupy a wide variety of habitats and environments and, just because they may, evolutionarily speaking, have the potential to digest cellulose efficiently, this does not of course mean that they have to specialise in this way. Rather, the evolutionary level of morphological development offers a flexibility, a greater degree of 'choice' as to what any given animal may feed upon. Some remain selective feeders on foodstuffs offering concentrated nutrients; others do specialise in satisfying their nutritional needs from fermentation of the cellulose of high-fibre forage. And, interestingly enough, the whole structure of the rumeno-reticular complex adapts to the type of diet taken. Those species which remain selective concentrate-feeders — for example roe deer, or the almost entirely frugivorous brocket deer (page 25) — show poor development of the fore-stomach, since they do not need a well-developed

Figure 3.3 The diet of the Himalayan musk deer (based on data from Green, 1985)

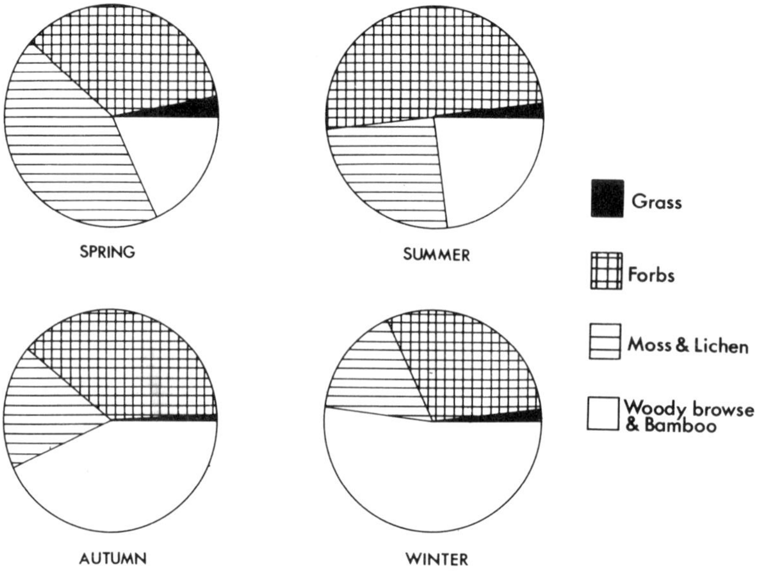

SPRING

SUMMER

AUTUMN

WINTER

Grass

Forbs

Moss & Lichen

Woody browse & Bamboo

rumen or reticulum; those on the other hand which rely on ruminal fermentation of cellulose have this region of the gut particularly well developed. Indeed, there is a whole suite of adaptations which accord with a particular type of diet: the size and relative proportion of the rumen and reticulum, and their relative size compared with the omasum and abomasum; the fine structure of the lining of the stomach and intestinal tract (reflecting the type of materials being absorbed); the actual species composition of the symbiotic micro-organisms within the rumen; and the retention time, the time the food stays within the gut as a whole. All these change, and change 'in sympathy', in adaptation to a particular type of diet; and, while the most striking differences are apparent in comparisons between different species with different foraging styles, similar changes, though on a finer scale, are apparent within species if feeding behaviour and diet are subject to seasonal variation.

Originally, deer were classified in terms of foraging behaviour into 'browsers', those which fed predominantly on woody materials and leaves of trees and shrubs, or 'grazers', those which fed primarily on grasses and herbaceous vegetation (with perhaps some recognition that some species occupied an intermediate position). Such a classification suffers, however, from being grossly oversimplistic. Roe deer, for example, in their 'typical' woodland habitat, feed largely on 'browse', plucking young leaves and new shoots from woodland trees and taking leaves of bramble, rose and other understorey shrubs (Drozdz, 1979; Henry, 1978; Jackson, 1980; Hosey, 1981). As we have seen, however, roe deer are also able to flourish in the open agricultural landscapes of Middle Europe, with little access to woodland or scrub for either shelter or forage. In such situations they are by no means browsers, for they feed extensively on cereal crops in their early stages of rapid growth, on broadleaved weeds among the crop as the cereal itself matures, and then return to the cereal itself as it ripens, plucking off the milky ears of wheat, barley and maize

Figure 3.4 Diet of muntjac deer in Britain (based on data from Harris and Forde, 1986)

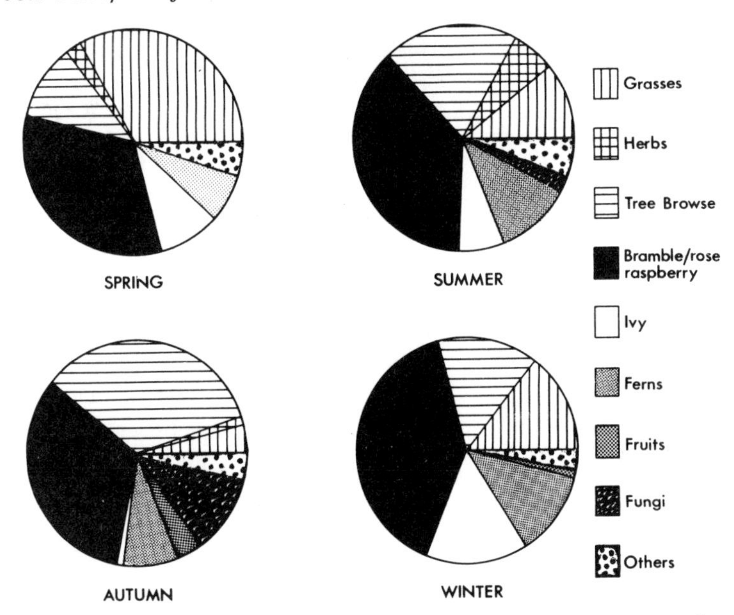

(Kaluzinski, 1982; Holisova *et al.*, 1984). While they are clearly not browsing in such instances, in both situations they are selectively taking from the environment small morsels of the most nutritious and concentrated foodstuffs. Classification of deer species as concentrate-selectors, bulk feeders, or intermediate (Hofmann, 1973) is thus a more meaningful way of describing their foraging habits, and such classification has the added advantage that it is more immediately relatable to gut structure and morphology. Indeed, one can determine the feeding style of any deer species from its gut structure (Hofmann, 1973; 1976; 1985), or the composition and activity of its ruminal symbionts (Hoppe *et al.*, 1977; de Bie, 1988).

It was Hofmann who, in 1973, first recognised that the gut structure of deer, and other ruminants, closely reflected feeding style in this way and could be used as a clear indicator of foraging strategy. On the basis of the absolute and relative size and degree of development of the rumen and reticulum, and the structure of the absorptive lining of the gut (which adapts itself differently to high-nitrogen and low-nitrogen diets), Hofmann recognised the three different categories we have just described, noting that concentrate-selectors like the roe deer tend to have relatively simple guts. The rumen is small, and evenly covered in short papillae; the opening to the reticulum is wide and retention time is short. Both omasum and abomasum are relatively small and underdeveloped and the intestine is short. The animals are relying heavily on cell contents and easily digestible parts of plants. By contrast, bulk feeders, which feed relatively unselectively and rely for their nutrient intake on relatively efficient digestion of cellulose, mostly have a large 'stomach' capacity overall, to cope with the large bulk of material they must ingest (and the fact that it must be held for longer within the gut for digestion to be effected). Specifically, too, the reticulum, that chamber where the cellulose is properly fermented, will be extremely large (and the more

Table 3.4
(a) Relative size of reticulo-rumen in concentrate-selectors, intermediate feeders and bulk feeders (from Harrington, 1985, and van de Veen, 1979)

Species	Body weight (kg)	Rumeno-reticular volume in relation to body weight (litres/100 kg)
Roe deer	14	8
White-tailed deer	39	10
Mule deer	57	10
Fallow deer	40	14
Red deer	95	23

(b) Relative length of intestine and ratio of small and large intestines in representative concentrate-selectors, intermediate feeders and bulk feeders (data from Hofmann, 1985)

Species	Total length of intestine as proportion of body length (BL)	Ratio in length of small:large intestine
Roe deer	12–15 × BL	70:30
White-tailed deer	12–15 × BL	70:30
Red deer	15–17 × BL	75:25
Fallow deer	15–17 × BL	78:22
Pure grazers	25–30 × BL	82:18

fibrous the forage on which the animal depends, the larger the relative volume of the reticulum). Intermediate feeders, as one might anticipate, show characters intermediate between these two extremes. Throughout, the main differences seem to be in relative size of the reticulum (or more properly rumeno-reticular complex), far larger in proportion to body size in the bulk fermenters, and in relative length of the intestine. Table 3.4 shows the relative volume of the reticulo-rumen and relative length of hind gut in a number of cervid species by way of illustration.

Hofmann's scheme was developed for ungulates in general, and was derived primarily from studies of antelope and other bovids. Within such a scheme, however, he has also attempted to classify (based purely on analysis of gut structure) various deer species (Figure 3.5): this anatomical classification matches exactly what is known of the ecology of these various species. Thus roe deer, as we would predict from our analysis of their feeding habits, show a gut structure, too, which would identify them as pure concentrate-selectors, as do musk deer, muntjac and brocket deer, species well known to select nutritious browse (Figures 3.3, 3.4). Most of these are relatively small species, but the largest of all living cervids, the moose, in spite of its enormous size, also shows anatomical features appropriate to a concentrate-selector and, as a selective feeder on aquatic marsh plants and browse materials, its feeding habits indeed match up to expectation. Species we know to be primarily grassland feeders, such as fallow deer (Figure 3.6) or barasingha,

Figure 3.5 Classification of the Cervidae on the basis of feeding style. The position of a species is based on structure of the digestive tract and on feeding behaviour/forage selection. Source: Hofmann (1985)

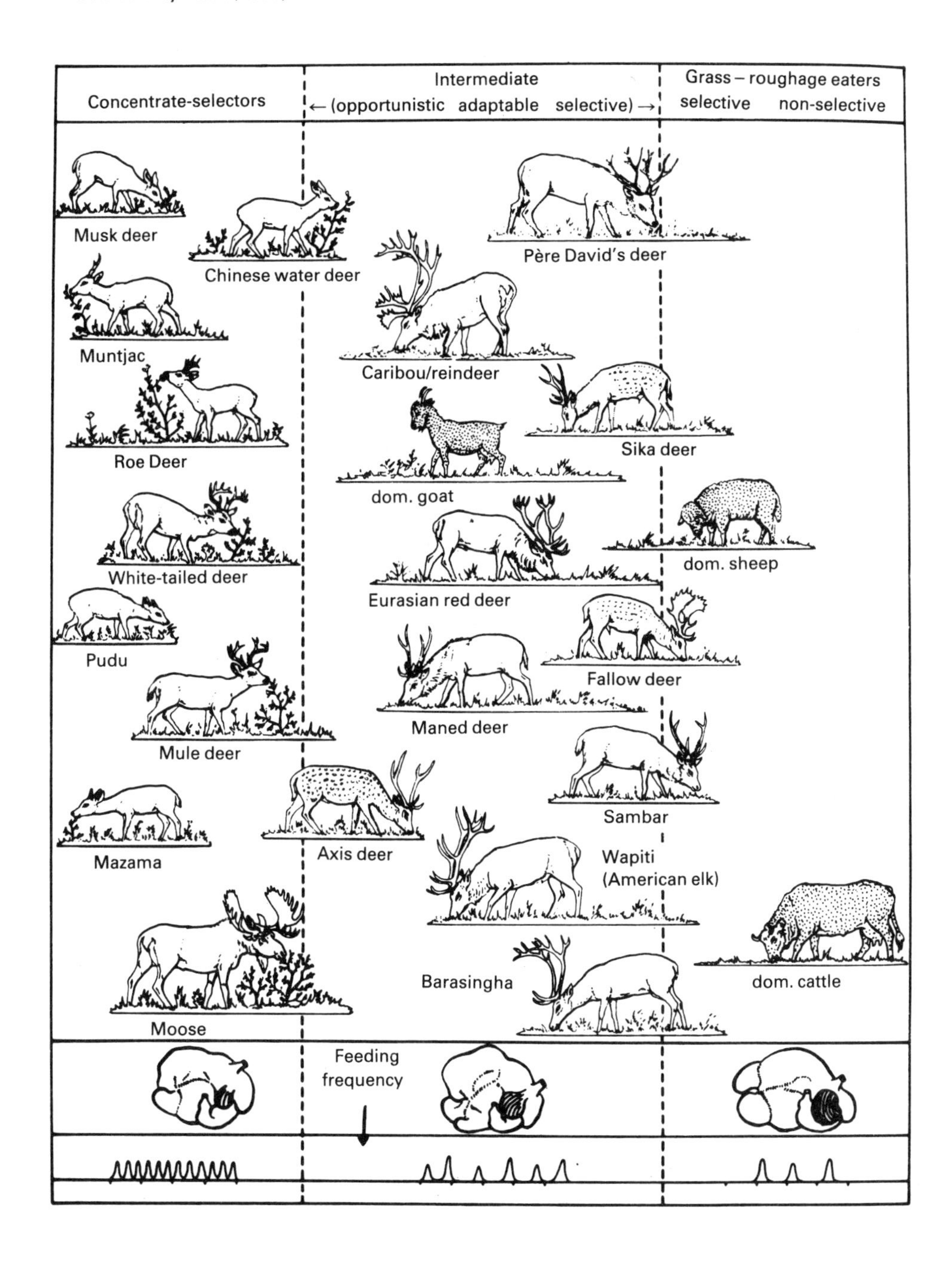

Figure 3.6 Diet of New Forest fallow deer (based on data from Jackson, 1977)

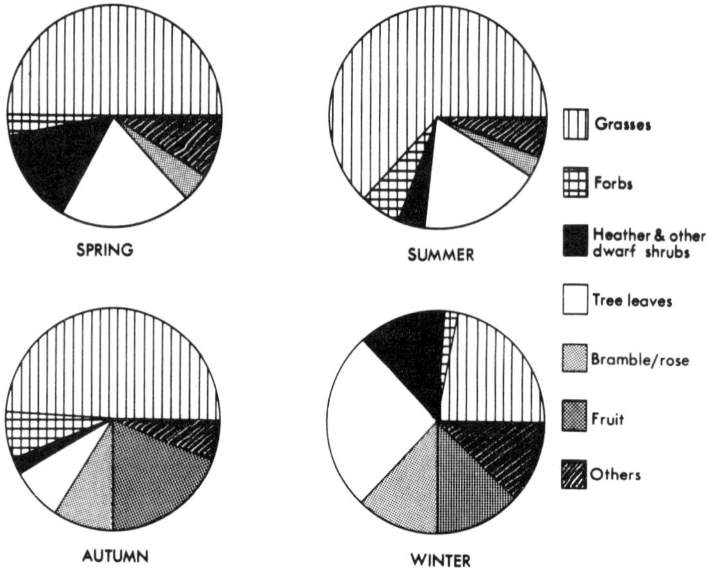

would also be identified in the same way on morphological criteria, and indeed the match throughout appears excellent.

Gut structure is not the only thing to change with diet. Species which have to feed selectively tend to have narrower heads and muzzles than do bulk grazers of equivalent body size, enabling them to select the more concentrated morsels more precisely (Figure 3.7; Gordon and Illius, 1988). Moreover, as already noted on page 44, even the gut flora adapt to diet. The high proportion of easily fermentable plant components in the diet of concentrate-selectors favours a microflora in the rumen composed primarily of amylolytic organisms, whereas cellulolytic microbes are found in higher numbers in the rumen of bulk and roughage feeders (de Bie, 1988). Intermediate feeders again resemble one or other category depending upon the precise diet taken. Each is precisely adapted to the type of plant material to be handled, and in fact the gut flora of concentrate-selectors, dominated by amylolytic species, will digest fibrous material but poorly, while the microflora in the rumeno-reticulum of a bulk feeder will not effect the best digestion of concentrated diets.

Steven de Bie was able to demonstrate this extremely clearly in an elegant series of experiments in which foodstuffs of different character and quality were incubated in rumen fluid taken from various ungulate species from West Africa which might be considered concentrate-selectors, bulk feeders or intermediate feeders (de Bie, 1988). Foodstuffs used were grasses (both old leaves and new growth), forbs and various browse materials, and de Bie's results showed clearly that the fastest rate of digestion of high-quality materials was accomplished by the rumen microflora of concentrate-selectors (in his experiments, bushbuck, *Tragelaphus scriptus*), while such materials were digested more slowly by ruminal flora of bulk or intermediate feeders (sheep and Zebu cattle). In complement, the gut micro-organisms of the bushbuck were unable to

Figure 3.7 Among animals of equivalent body size, muzzle width and shape of the incisor arcade (bite) also correlate with feeding style. The figure shows the muzzle and 'bite shape' of (a) a grazer, (b) an intermediate feeder, and (c) a browser of equivalent body weight (c.250 kg). Source: Gordon and Illius (1988).

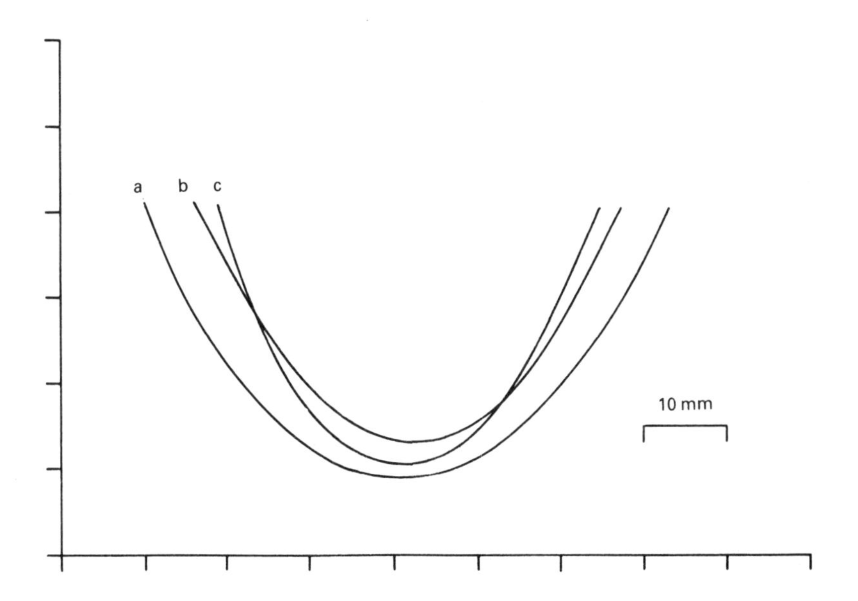

make any impression at all on the high-fibre substrate provided by grass forage; yet grasses were digested fairly rapidly by sheep and goats. This match between the composition and character of the ruminal microflora and the type of diet normally taken by the host is indeed so precise that de Bie suggests that his experiments may even offer a method whereby the dietary habits may be established of animals whose feeding ecology is otherwise poorly understood: if samples of vegetation ranging from low-quality fibre-rich materials to high-quality low-fibre plants are incubated in rumen fluid taken from the species concerned, the level of digestive activity recorded for each sample can give good information about the type of diet to which the animal is normally adapted (de Bie, 1988).

Finally, although all these changes are shown in adaptation to the different foraging styles of different species, even within a species seasonal changes in relative size of rumen and reticulum, changes in the gut lining and changes in rumen micro-organisms accompany seasonal changes in foraging style, where an animal must adopt a different type of diet at different times of year (Hofmann and Schnorr, 1982; Hofmann, 1985). Indeed, as a final twist, the relationship between body size, rumeno-reticular volume, mouthpart size and diet may even be reversed. We noted in Chapter 2 that, particularly in polygynous deer species, a marked size difference may exist between the sexes, with males considerably larger and heavier than females. This may be imposed by sexual competition (page 29) and not itself be in response to diet, but may, conversely, impose a dietary difference between the sexes, as each forages for the diet appropriate to its size, mouthparts and ruminal volume. Thus Staines *et al.* (1982) have shown a marked difference in diet, both in actual composition of the diet and in nutritional quality, between male and female red deer in Scotland. Compared with stags, hinds ate less heather and more grass, particularly the more digestible fine-leaved species. The differences in plant species taken meant that, overall, hinds were feeding on plants which offered higher proportions of nitrogen (as grams per 100 g dry matter) than did those fed on by stags. The total amount of available nitrogen contained in the food at any one time was, however, similar in both sexes, since the rumen volume of stags was greater than that of hinds. To some extent the difference in diet between the sexes is due to the fact that (in turn owing to a degree of sexual segregation) stags and hinds may occupy separate geographical ranges, where availability of different foodstuffs may not be the same; but, even in areas of range overlap, the differences in species composition and quality of the diet were retained. It would appear therefore that stags and hinds are actually selecting different foods, each adopting the foraging strategy appropriate to their body size, mouthpart size and ruminal physiology. It is even possible that these differences in the way the two sexes must satisfy their nutritional requirements — because of their differences in body size and physiology — are in themselves a major cause of the sexual segregation in the first place (see page 62).

Seasonal Changes in Composition of the Diet

Availability and quality of particular forages will differ at different times of year: between winter and summer ranges for those species or populations which undertake a seasonal migration between distinctly different geographic areas (as do some populations of reindeer for example); or between wet and dry seasons or summer and winter, even for residents in tropical or temperate areas. In the same way that patterns of habitat use must change over time to satisfy requirements for cover, shelter and forage as the facilities offered by given habitats themselves vary, so, within the deer's 'gross' foraging style of concentrate-selector, intermediate feeder or bulk feeder, the actual species composition of plants selected must change to accommodate changes in availability and quality of given forages.

This is clearly forced upon animals such as reindeer, which occupy strikingly different environments at two ends of a seasonal migration. Reindeer on the tundra of the high Arctic feed through the brief summer to a large extent on lichens and mosses; in their woodland winter ranges they browse far more extensively on woody species and broadleaved herbs. Even for species which do not undertake long migrations of this kind to totally different biotopes, however, those for whom perhaps the same array of plant species may remain available throughout the year, the relative abundance or food quality of each may vary considerably. (The nutritional value of a given species of grass, for example, is clearly going to be far higher during the season of its growth than when the growing season is complete; during the latter period the grass may still be available, but will be dry and unpalatable.)

Seasonal variation in actual diet, in response to this changing availability and quality of forage, may thus be very marked even among the more sedentary species of deer. The diet of barasingha in one study in central India (Martin, 1978), for example, consisted during the dry season (December–May) predominantly of only three species of grasses: *Saccharum spontaneum, Bothriocloa odorata* and *Themeda triandra.* By contrast, during the monsoon growing season, a huge variety of species was fed upon, as each in turn provided an abundance of new green shoots. Preferences for particular species over others were also far less conspicuous during the monsoon than during the dry season: *Bothriocloa odorata* was still an important component of the diet, but less so than through the dry season, and *Saccharum spontaneum*, the dry-season staple, was scarcely eaten.

The diet of chital in Sri Lanka shows even more marked seasonal change in composition, again reflecting variation in availability and quality of different forages. Table 3.5 shows the plants species on which chital were observed to feed in Ruhuna National Park (Balasubramaniam *et al.*, 1980). Of a total of 36 species included in the diet as a whole, only seven were taken in both wet and dry seasons; in addition, there was a distinct shift from feeding on grasses and herbs when these were available in the wet season, to a diet composed predominantly of shrubby or woody species during the dry season, when the preferred herbaceous species were presumably unavailable.

In such environments as these, the rainy season is of course the period of plant growth; in more temperate climes the growing season is dictated by temperature more than by rainfall, and contrasts in availability and quality of food will tend to be between summer and winter. Dietary composition again reflects this, and other more subtle changes in relative quality or availability of different foodstuffs. Diets of fallow deer, roe and sika in the New Forest show clear contrast in species selected between summer and winter, but also show gradual change even from month to month (Putman, 1986a, 1986b) (Tables 3.6, 3.7, 3.8). Table 3.6, for example, based on Jackson's studies of the diet of New Forest fallow deer,

Table 3.5 Seasonal changes in the diet preference of the spotted deer or chital in the Ruhuna National Park, Sri Lanka (from Balasubramaniam *et al.*, 1980)

Category	Species	Family	Season	
			dry	wet
Grass	*Eragrostris viscosa*	Gramineae	+	+
	Dactylotaenium aegyptium	Gramineae	+	+
	Sporobolus diandrus	Gramineae	+	+
	Alloteropsis cimicina	Gramineae	+	+
	Echinochloa colonum	Gramineae		+
	Setaria pallidifusca	Gramineae		+
	Cyanodon dactylon	Gramineae		+
	Panicum sp.	Gramineae		+
	Auxonopus sp.	Gramineae		+
Sedge	*Cyperus iria*	Cyperaceae		+
Legume	*Tephrosia purpurea*	Leguminosae		+
	Zornia diphylla	Leguminosae		+
	Desmodium triflorum	Leguminosae		+
Herb	*Spermococe hispida*	Rubiaceae		+
	Corchorus tridens	Tiliaceae		+
	Cyanotis axillaris	Commelinaceae		+
	Aneilema spiratum	Commelinaceae		+
	Rungia repens	Acanthaceae		+
Thorny shrub	*Flueggea leucopyrus*	Euphorbiaceae	+	
	Capparis sepiaria	Euphorbiaceae	+	
	Azima tetracantha	Salvadoraceae	+	
	Carissa spinarum	Apocyanaceae	+	
Shrub	*Lantana camara*	Verbenaceae		+
	Memecylon umbellatum	Melastomataceae	+	
	Cassia auriculata	Leguminosae	+	
Tree	*Salvadora persica*	Salvadoraceae	+	+
	Crateva religiosa	Capparidaceae	+	+
	Capparis zeylanica	Capparidaceae	+	
	Feronia limonia	Rutaceae	+	+
	Atalantia monophylla	Rutaceae	+	
	Drypetes sepiaria	Euphorbiaceae	+	
	Euphorbia antiquorum	Euphorbiaceae	+	
	Mischodon zeylanica	Euphorbiaceae	+	
	Sapindus emarginatus	Sapindaceae	+	
	Aglaia roxburghiana	Meliaceae	+	
	Terminalia arjuna	Combretaceae	+	

shows that throughout the year the deer are primarily grazers (as their gut morphology might suggest: Figure 3.5). Through the growing season, from March to September, grasses form the principal food, comprising in the region of 60 per cent of total food intake, with herbs and broadleaf browse also making a significant contribution. Acorns and other tree fruits are a characteristic food through autumn and early winter, although their importance in the diet varies from year to year with variations in the mast crop. Other major foods through the autumn and winter, which become increasingly important as the year's mast is exhausted, are bramble, holly, ivy, heather and browse from felled conifers. Even at this time, however, grass still makes up in excess of 20 per cent of the diet.

Table 3.6 Monthly variation in the composition of the diet of fallow deer in the New Forest, England. Percentage composition of the diet in each month is calculated from data of Jackson (1977)

Jan	Feb	Mar	Apr	May/Jun/Jul	Aug	Sep	Oct	Nov	Dec	
21	25	59	67	63	57	58	33	25	21	Grasses
1	1	1	6	6	12	7	2	2	1	Herbs
14	14	7	1	0	0	0	0	8	17	Conifers
12	17	9	7	4	3	1	0	2	7	Holly
26	12	3	9	21	24	15	14	26	16	Other broadleaves
16	24	16	3	4	3	2	1	8	16	Heather
0	0	0	0	0	0	0	0	0	0	Gorse
2	0	0	0	0	0	14	41	22	15	Fruits
8	7	5	7	2	1	3	9	7	7	Other

It is evident from this that the deer are preferential grazers throughout the year and take increasing amounts of browse and other foods through autumn and winter merely to compensate for lack of grazing materials outside the growing season. The changes are, however, marked, and differences in the composition of the diet are actually statistically significant even from month to month (Putman, 1986). Similar differences are seen in New Forest sika (Mann, 1983) and roe (Jackson, 1980), whose diets are shown in Tables 3.7 and 3.8 for comparison, and indeed in almost any temperate species which have been studied.

Table 3.7 Monthly variation in the composition of the diet of sika deer in the New Forest, England. Percentage composition of the diet in each month is calculated from data of Mann (1983)

Jan	Feb	Mar	Apr	May	Jun	Jul	Aug	Sep	Oct	Nov	Dec	
25	25	22	39	38	40	39	50	44	31	28	27	Grasses
0	0	0	0	0	0	0	0	0	0	0	0	Herbs
20	19	23	13	2	0	0	1	0	6	8	16	Conifers
1	3	1	1	1	1	1	1	2	2	1	1	Holly
11	11	10	13	14	14	16	10	19	25	14	14	Other broadleaves
24	23	30	23	35	37	35	29	27	23	24	25	Heather
7	14	8	7	6	5	6	6	7	4	7	5	Gorse
6	4	2	0	0	0	0	0	0	6	14	9	Fruits
6	1	4	4	4	3	3	3	1	3	4	3	Other

What, then, determines the diet selected? What determines which particular species a given animal will feed on at any given time? To some extent this will be dictated by availability: the striking contrast between winter and summer diets of migrating caribou in North America is clearly due in large part to the fact that lichens are the most abundant food source available to them in the high tundra where they spend the summer, while totally different foodstuffs await them in their winter range. At a finer scale, availability of different foodstuffs will change around the year even within a more restricted home range, and here, too, as we have seen, nutritional quality of different forages is also changing.

Table 3.8 Monthly variation in the composition of the diet of roe deer in the New Forest, England. Percentage composition of the diet in each month is calculated from data of Jackson (1980)

Jan	Feb	Mar	Apr	May	Jun	Jul	Aug	Sep	Oct	Nov	Dec	
4	5	5	10	7	8	8	8	8	9	10	4	Grasses
5	2	2	30	16	16	16	17	17	4	4	6	Herbs
33	22	22	5	1	8	0	0	0	12	12	13	Conifers
0	2	0	0	0	14	1	0	0	0	0	0	Holly
45	52	52	41	71	50	58	58	55	54	53	50	Other broadleaves
6	14	14	14	5	4	7	7	5	5	4	7	Heather
0	0	0	0	0	0	0	0	0	0	0	0	Gorse
1	1	1	0	0	0	0	0	8	8	17	7	Fruits
6	2	3	0	0	0	10	10	7	8	0	13	Other

Clearly, any individual animal will want to optimise both quantity and quality of its food intake at any given time; the diet selected will thus be a compromise between preference for the most nutritious species and the relative availability of these species against other, perhaps less preferred but more abundant foodstuffs. Actual food preferences have sometimes been established for captive individuals by offering them choices of different natural foodstuffs in feeding trials (e.g. Heady, 1964; Radwan and Crouch, 1974). Rarely do the preferences expressed reflect exactly the actual diet recorded in the wild, for preference in isolation must always be traded off against availability in practice.

Yet diets do change, and the animals are obviously selecting for something. Analyses of the relative species composition of the diets of animals in the wild (and, perhaps more illuminating, comparisons of the relative quality of the same plant species at times when they are included in the diet and times when they are not) suggest that most deer select from among the foods available those which are at any given time highest in protein, and best in terms of digestibility (proportion of fibre, and, of that fibre, proportion made up of relatively easily digested forms of cellulose). Thus, for example, when diet of fallow deer in southern England (Parfitt in Putman, 1986) was compared with available standing crop, energy value, digestibility, and content of nitrogen, calcium, potassium, phosphorus and magnesium, the one consistent relationship that emerged was with digestible nitrogen. Over the winter months the deer appeared to select those forages highest in available protein. Interestingly enough, this correlation between nitrogen content of forages and their importance in the diet held true only between November and March. During the summer, growing season, when perhaps all types of forage offer adequate protein and there is not the same need to select so carefully, the relationship breaks down. Diets selected by wapiti and black-tailed deer were studied by Hanley (1984) in Washington, USA. Both species are essentially intermediate feeders on Hofmann's (1973) classification, but wapiti are perhaps more towards the bulk-feeding end of the scale while black-tailed deer are relatively more selective feeders. Diets selected by the two species thus differed, with forbs and woody browse constituting a higher proportion of the diet of black-tailed deer and grasses contributing more to the diet of the wapiti, and with these differences clearly imposed upon them by their respective gut structures. Within their different dietary 'styles', however, both deer were shown to be selecting, from among those plant species available to them, those which offered at any season the highest concentrations of cell solubles (black-tailed deer) and highest proportion of digestible cellulose coupled with high cell solubles (wapiti). Similar responses have been shown in a variety of other species.

Cycles of Appetite

One other curiosity we should perhaps note here is that deer of seasonally-changing environments not only show variation in the species composition of the food they take at different times, but also show marked changes in appetite. Such changes in food intake are accompanied by cycles in body weight and condition, and such cycles have been reported for white-tailed deer (McEwen *et al.*, 1959; Cowan and Long, 1962; Silver *et al.*, 1969; Short *et al.*, 1969; Holter *et al.*, 1977) and black-tailed deer (Wood *et al.*, 1962), red deer (e.g. Kay, 1979; 1985; Kay and Staines, 1981; Fennessy, 1981) and a number of other species (moose: Gasaway and Coady, 1974; reindeer: McEwan and Whitehead, 1970; and roe deer: Drozdz *et al.*, 1975). It might be argued that the loss of body weight/condition and the decline in food intake over winter among these temperate species was due to a reduced quality or availability of forage; but we have already seen that deer adapt their diet in different seasons to take advantage of what foodstuffs are available to them, and are capable of selecting from among them those of the best quality. Further, these cycles of body weight and appetite are recorded even for captive animals kept at constant temperature, and offered *ad lib.* food, but subject to the changing daylight regimes of the temperate year. Clearly, the reduction in appetite and loss of body weight over winter is a purely endogenous cycle and not imposed by environmental strictures. The whole syndrome is in fact also accompanied by a decrease in basal metabolic rate in the winter and by a tendency to reduce activity.

The phenomenon is hard to explain. Despite the apparent abundance of winter foods available to deer in the wild, are the animals in fact limited to what they can ingest: to the extent that they cannot perhaps meet the daily energy demands of normal activity, increased still further by the colder temperatures of winter? Do metabolic rate and activity drop in an attempt to minimise the consequences of this imbalance between restricted income and increased expenditure of energy? And is it because such adjustments cannot compensate completely for the shortfall, and thus the animal must draw on bodily reserves of energy, that a decline in body weight is recorded? It is possible that, although deer are able to switch their feeding patterns to take in forages which are readily available in winter and their guts remain full, a decrease in the quality of the food available overall will lead to greater reliance on more fibrous foods, which will take longer to digest and will require a longer retention time within the gut. This in turn will mean that the rumen empties more slowly and the animal will be able to fill it less often: not only causing the observed apparent loss of appetite but also, because of a reduced food intake and an intake composed of food of lower quality, a reduced total energy intake. Cycles of food intake, body weight and metabolic rate are, however, still apparent in animals kept at constant temperature and offered superabundant food of constant quality, suggesting that the syndrome is an evolutionarily determined endogenous cycle rather than an immediate and direct response to lowered food quality and limiting gut fill. Is the whole relationship the other way around? Does the animal first voluntarily reduce activity and lower basal metabolic rate to conserve energy and reduce heat losses over winter, with lowered food intake and loss of body weight consequent upon that? The major imbalance in the winter energy budget may, after all, be the increased cost of energy lost through chilling. If an animal lies down, and becomes inactive, heat loss is reduced — and indeed only the upper part of the body surface is available as a heat exchanger with the 'outside air'. When standing to forage, heat is lost from the entire body surface, and it is economical to stand only if energy intake through this foraging activity exceeds the energy loss. If in response to this the animal becomes relatively inactive in winter, however, although it does indeed reduce its losses it may end up hardly feeding at all. Again it must meet its energy demands, however reduced, from somewhere, so it must draw on its reserves and therefore body weight falls.

This is, though, a real chicken-and-egg of a problem and no-one really knows whether metabolic rate falls as a result of reduced energy intake, or whether intake falls as a result of reduced activity. In laboratory experiments both changes are very closely synchronised. It is even possible that we are wrong to think of the phenomenon as a reduction of appetite and metabolic rate to below normal in the winter. Could it not be that the winter situation is the norm and that food intake, metabolic rate and body-weight gain are *above* normal in the summer? Here, metabolic rate may increase through increased activity associated with setting up a territory, capturing a harem, mating etc. (for males), or pregnancy and lactation (for females). Energy requirements are thus high but easily met from abundant and relatively more digestible summer foodstuffs. Indeed, these are so abundant and so digestible that the animal can ingest far more than it actually requires, develops a positive energy surplus, and puts on body weight. This is perhaps an equally plausible explanation for the changes observed.

USE OF TIME

Limitations of gut fill may not only restrict total feed intake, but may also influence feeding patterns in the shorter term. When the rumen is full, ingestion must by definition cease and rumination can commence; indeed, it is thought to be the actual sensation of a full rumen that directly stimulates regurgitation and triggers off a period of rumination. In consequence, many species of deer show clear patterns in activity of alternating periods of feeding and of cudding. If the animals are undisturbed, this feeding rhythm resolves itself with a periodicity dependent on feeding rate and rumen volume (but usually in the region of two to three hours). Jackson (1974), for example, showed that fallow deer in the New Forest of southern England tended to feed for a period of three to four hours, and then to rest and cud for two to three hours, and that this rhythm was maintained throughout the day.

Figure 3.8 The activity pattern of one individual male Himalayan musk deer on three different nights. Source: Green (1985)

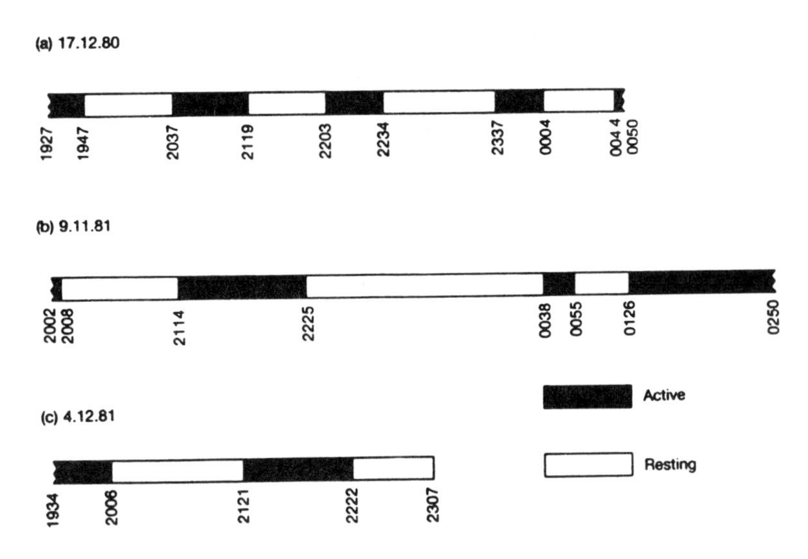

Whether or not it is determined by feeding activity or rumen fill, most deer species do show clear activity rhythms of this sort. Fallow deer are bulk feeders. Looking by way of contrast at a clear concentrate-selector, Figure 3.8 shows the activity pattern of one individual male Himalayan musk deer on three different nights (Green, 1985), and it is clear that there is a very similar pattern of alternating bouts of feeding activity (on average, of about 60 minutes) and resting/rumination (on average 50 minutes).

It is, however, somewhat unusual for such a regular pattern to be maintained throughout the 24 hours, for, where animals are subject to disturbance (either by humans or by other potential predators) more at certain times of day than at others, activity patterns may become shifted to favour quieter periods, or may retain their regularity, but become restricted within a shorter period of activity overall. Green's musk deer, for example, were effectively nocturnal in habit, and while, during the hours of darkness, they maintained the periodicity suggested by Figure 3.8, during daylight feeding bouts were suppressed completely and the animals remained in cover, resting and ruminating throughout. A number of species or populations respond in this way, becoming effectively nocturnal, or most active perhaps at dawn and dusk (as, for example, barasingha in Martin's (1978) study in central India); others are more daylight-active (perhaps because, for them, risk of predation is greater at night).

It seems clear, however, that this restriction of activity to certain periods of the day is in response to disturbance or risk of predation, and that all species could potentially maintain the regular alternation of feeding/rumination throughout the 24 hours. Comparison of the behaviour patterns of populations of the same species in disturbed and relatively less disturbed environments would seem to support this: van de Veen (1976) found that red deer in the Netherlands could be observed

Table 3.9 Activity rhythms of New Forest sika deer: figures show, for each month, the proportion of all sightings seen at different times of day (data from Mann, 1983)

Hours	Jan	Feb	Mar	Apr	May	Jun	Jul	Aug	Sep	Oct	Nov	Dec
0—02.00	9	5	7	2	7	7	7	9	10	8	10	2
02.00—04.00	9	5	3	2	7	3	5	7	5	7	9	3
04.00—06.00	5	2	2	8	10	10	13	11	4	3	5	6
06.00—08.00	4	5	3	3	11	10	9	9	3	5	4	6
08.00—10.00	11	5	2	15	6	9	5	5	2	13	15	12
10.00—12.00	11	13	23	4	9	7	4	3	2	8	4	8
12.00—14.00	4	4	12	4	5	7	2	1	5	4	2	6
14.00—16.00	2	5	9	7	8	5	2	2	3	9	6	5
16.00—18.00	20	16	9	20	8	6	2	6	13	12	14	17
18.00—20.00	10	16	18	23	7	8	13	10	21	14	9	15
20.00—22.00	7	12	3	10	18	16	22	24	19	10	6	10
22.00—24.00	8	9	9	2	3	11	16	13	12	7	16	10

active at any time of day in forests where they were subject to a low level of disturbance, but in disturbed areas activity became restricted to brief periods around dawn and dusk. Sika deer in the New Forest in England were again almost equally likely to be seen active at any time during the 24 hours (Mann, 1983; Table 3.9); in the neighbouring forests of the Poole Basin in Dorset, the same species is strictly nocturnal. The New Forest is vegetationally diverse and presents a mosaic of suitable feeding areas and habitats offering cover in a fine-grained mixture. The sika population can feed and rest within the body of the Forest itself, effectively undisturbed. The coniferous forests of Wareham and Purbeck in the Poole Basin provide excellent cover, but little food, so the sika must leave the forest to feed in agricultural fields in the surrounding area; here they are of course far more exposed, and feeding activity is restricted to the hours of darkness.

Similar effects have been observed in other species: white-tailed deer, for example, have been shown to alter their activity rhythms in response to both human disturbance (Montgomery, 1963; Marchinton, 1964; Kammermeyer, 1975) and activity of predators (Jackson *et al.*, 1972). Further evidence that it is disturbance or risk of predation that causes so many populations to restrict activity to particular periods can also be taken from the fact that, among those species which have adopted effectively nocturnal habits, it is notable that they tend not to come out even at night if it is very clear, or if there is a full moon, when they would of course be more clearly visible to predators. Quality of light is of no import to the deer themselves, since they are not essentially visual animals. Since they rely more heavily on sound and smell, their ability to detect predators is no better or worse on clear or overcast nights; they will, however, be more apparent to the predator on clear nights.

All the indications are that deer are, so to speak, physiologically, 24-hour animals, with regular activity rhythms potentially maintained throughout, and that curtailment of this rhythm, or suppression of activity at certain periods, is in relation to disturbance or risk of predation.

4 Social organisation and behaviour

Evolution within the Cervidae appears to be accompanied in general by an increase in size (page 30). In addition it would appear that there is also a tendency for increasing sociality, for, while musk deer, water deer and muntjacs are essentially rather shy and solitary animals, many of the more advanced Cervinae or Odocoilinae are frequently encountered in quite large herds (e.g. Geist, 1974; Kurt, 1978). Group sizes even in these more social species, however, change with season, with habitat — with need. The groups formed by many species are fluid, changing in both size and composition as individuals join them, stay for a while and then move on. Indeed, it may be more accurate to suggest that evolution has been accompanied by a decrease in anti-sociality rather than an increase in sociality itself, for the groups are often far from permanent and appear to be formed almost for convenience in response to defined short-term requirements. Even among those species where larger groups may be encountered, the basic unit is still very much the individual: a single male, or female, or a female and dependent offspring (Putman, 1981).

At this stage it is important to define our terms more carefully, for there is a clear distinction between strict, fundamental social groupings — what we may properly call the social unit — and casual associations of more than one of these units to form larger groups. Thus, while among fallow deer, to take an example, herds of between 70 and 100 animals are regularly observed on favoured feeding grounds, closer observation will reveal that this is in fact a coincidental aggregation of a number of separate smaller social units, occurring together merely because they happen to be using the same area simultaneously. If such a 'herd' is observed for a long period, it may be seen that it is not of constant constitution: small groups may be seen to join the aggregation, and others to leave. In effect, we must recognise that there are two levels of social organisation: that the feeding groups are no more than casual associations of a variety of sub-units is clear from the fact that, as the group disperses from the feeding grounds, it fragments once more into groups of up to seven or eight individuals, moving off in separate directions.

The same pattern of an association of a number of essentially small social units into larger feeding groups is seen in all other 'social species', and the fact that these smaller social units are not just temporary sub-units of a true and constant herd, but that they are themselves the basic unit, with the feeding herd formed as a chance association of the individual units, is clear from the work of Horwood and Masters (1970) on a marked population of sika. Any one marked hind might be seen on the feeding grounds in company with any other marked hind in random combination: there was no constancy of herd structure or association between individuals.

These smaller groupings may, then, be considered the more fundamental social unit. It would appear, however, that in most cases these associations are little more stable than the larger aggregations, that even these groups are not persistent, nor constant in composition. Among fallow, these social units are normally of between three and seven or (exceptionally) up to 14 animals and are indeed the same sort of size of group in which fallow are more regularly encountered in most habitats (e.g. Jackson, 1974; Parfitt in Putman, 1986), again suggesting that they represent the basic unit. Despite the stable appearance of these groups in terms of number, however, it appears that the size of group reflects a general response to habitat (see pages 63–70), not the cohesiveness of social unit. In practice, the groups vary in individual composition from day to day, despite their constancy of size (e.g. Putman, 1981; Waterfield, 1986). The fundamental unit is thus far smaller than is often appreciated from casual observation, and it is suggested (Putman, 1981) that perhaps in *all* deer species the basic social unit is in fact extremely small — the family group, or associating pairs of males — with the 'herding' or 'non-herding' element depending upon the degree to which the separate units associate to feed.

This is clearly an oversimplification, and among the sika observed by Horwood and Masters, and from my own observations on fallow deer, it would appear that there is some restraint upon these associations. While a sika hind with a blue and yellow collar may on one night feed in company with a hind with a red and yellow collar, and on another night may, with her calf, be seen with a hind with a green collar (while red and yellow is some half a mile away with red and black), there will be certain animals with which she never feeds. Similarly, among fallow, there seems to be a limited 'pool' of herd-units which will associate with each other — separate and not interacting with an adjacent set of herd-units. While group A may be seen feeding with all or any of a set of groups B–K, they never associate with groups L–P, or vice versa. Since the animals of a given area of forest will tend to use the same feeding grounds and their total range is relatively small, it is of course inevitable that one is likely to see a relatively small set of animals together fairly frequently and that they will not interact with a set of social units whose ranges are polarised on a totally different set of feeding grounds.

Further, it is clear that in some species, or some populations of some species, the smaller groupings of up to perhaps a dozen animals are more persistent and stable in individual composition than in others. Some 'social' species, some populations, are undoubtedly as individualistic as suggested here, associating only temporarily in groups in response to environmental circumstance. In other cases, more stable association can be found: female red deer studied by Clutton-Brock and others on the Isle of Rhum, Scotland, formed distinct matrilineal groups which persisted for long periods.

Observations such as these led Kurt (1978) to recognise distinct types of sociality among those of the deer that do form groups at all (he developed his ideas primarily for Asian species, but the general principles in fact hold good for all). Thus he distinguished between species such as sambar and perhaps sika which, while essentially solitary, may aggregate temporarily for feeding and those species (Kurt cites chital as an example, and we might now include fallow or the Rhum red deer) in which the aggregations are somewhat longer-lasting and persist long enough that the individuals within them no longer remain entirely individualistic but begin to adopt some of the behaviour appropriate to group life, such as the formation of dominance hierarchies etc. (page 82).

If we weave such an idea into our overall scheme, then we may resolve the evolutionary development of sociality within the deer as a shift from solitary and territorial species such as the musk deer and muntjacs, through a stage, now represented perhaps by sambar and others, which are basically solitary but do form occasional aggregations, to a final, truly social level of more regular association at a stage where hierarchies are developed within a relatively persistent group (chital, ?fallow).

SEXUAL SEGREGATION

Clutton-Brock *et al.* found in the red deer of Rhum persistent matrilineal groups. This in itself introduces another rather curious characteristic of social organisation among the deer: in both solitary and more social species the two sexes operate almost completely independently of each other. While sexual separation among solitary animals is perhaps not surprising, even among the more social species the sexes remain separate for most of the year and come together only during the short breeding season to mate. For the rest of the year males and females exist in discrete single-sexed subpopulations: groups of mature males, and other groups of females and their offspring of perhaps the current and immediately preceding year. In many species, the sexes may even be geographically separated, with males and females occupying distinct and non-overlapping geographical ranges.

The actual degree of sexual segregation varies somewhat among species and among populations. As we shall see later, various characteristics of the habitat occupied may have an effect both on group size and on the degree of separation observed between the sexes. In many Scottish populations of red deer, for example, over 90 per cent of the animals seen are in segregated parties (Jackes, 1973; Mitchell *et al.*, 1977); but in one study of red deer in the Crimea only 18–30 per cent of the stags and some of the hinds were seen in segregated groups (Yanushko, 1957). Similar variation in 'rigidity' of sexual segregation has been reported for sika deer (Takatsuki, 1987) and fallow deer (Schaal, 1982; Waterfield, 1986). Overall, however, some level of segregation may be observed in most species.

The reasons for such sexual segregation are obscure, but it also occurs in sheep, goats and several other bovids. A number of possible theories have been put forward. Geist and Bromley (1978) have, for example, suggested that segregation of the sexes at least among the deer may have developed as an anti-predation strategy, with antlered males (furnished, through the possession of antlers, with some sort of weaponry as defence against predators) further reducing risk of predation by distancing themselves from the more vulnerable and defenceless 'small-deer'. Such a strategy would not be entirely selfish, since, by reducing both density and herd size in a given area, it would by the same token make any group, of either males or females, less obvious to local predators. Other theories relate more to observed differences in feeding strategies of the two sexes and are perhaps more universally applicable, among both deer and bovids.

As we have already noted, the increased tendency towards polygyny among more social species will lead to increased male competition, and this in turn is seen to be accompanied by an increase in size dimorphism. Increased body size, with the accompanying higher food requirements, may result in males having to feed on different forage species, or having to change their foraging style more towards bulk than concentrated feeding. Differences in food species taken, or in absolute bulk required, may force the sexes to forage in different areas differing in species composition of absolute available biomass and thus result in the sexual segregation observed.* Such a theory has been suggested for observed sexual segregation in red deer (e.g. Watson and Staines, 1978; Staines *et al.*, 1982; Clutton-Brock *et al.*, 1982) and may be applicable in other species, too.

Such segregation may be exaggerated by competition between the sexes. For, in principle, while the requirement for an increased food intake might by itself lead males to concentrate on different, perhaps more nutritious, plant species, or to capitalise on their greater ability to digest even poorer-quality foods by turning more towards bulk feeding, they could equally well satisfy this increased nutrient requirement merely by increasing intake of the same foodstuffs as selected by females: merely, in effect, by eating more of the same. So what causes the shift in dietary composition and habitat? What forces them to 'opt for' exploitation of their greater capacity to digest bulk foods? Geist and Petocz (1977) suggested that males might avoid habitats and feeding areas favoured by females in order to reduce the amount of competition they would themselves then impose on females which might carry their offspring, or even on those offspring themselves. This, however, suggests some degree of altruism which could evolve only through some form of group selection, since it cannot otherwise explain why non-breeding males also occupy ranges different from those of the females. Clutton-Brock and Harvey (1983) argued instead that larger-bodied animals might actually be at competitive disadvantage when co-occurring with smaller individuals (of the same or different species).

Assuming that bite size of any animal, and thus its food intake, are determined at least in part by incisor breadth, and observing that bite size does not change in simple linear relation to changes in body size or

---

*This switch towards a more bulk-feeding habit is facilitated by the fact that, physiologically, large-bodied animals are more capable of tolerating poorer-quality foods than are small animals (Jarman, 1974; Demment and van Soest, 1985).

nutrient requirements, Clutton-Brock and Harvey argued that large animals will not be able to maintain adequate food intake on swards of a height that would easily support smaller animals, and will thus be excluded even from habitat types preferred by both. Nutrient requirements of animals are in general related to the ¾ power of body size ($W^{0.75}$), while rumen volume or gut capacity are isometric with body weight. Incisor breadth, being a linear measure, not volumetric or three-dimensional as would be body volume or body mass, scales as $W^{0.33}$. If food intake is a function of bite area and sward depth (the two together effectively contributing to a mouthful of particular volume), then, as body size and metabolic needs increase, the increase in the rate of food intake, on a sward of given height, cannot keep pace (since bite size increases in relation to $W^{0.33}$, while metabolic needs scale to $W^{0.75}$). To maintain intake the animal must increasingly shift to swards of greater and greater depth, and in consequence large-bodied animals cannot maintain themselves on swards which would adequately support smaller beasts; thus competition will lead to exclusion of larger individuals even from preferred sward types.

Such an argument has been used primarily to explain niche differences between species; but it may also be invoked in explanation of differences in habitats and diets selected by different sexes of a single species, as we do here (and see also Clutton-Brock, Iason and Guinness, 1987). In a rather more formal analysis than we have attempted here, Illius and Gordon (1987) have demonstrated that weight differences of greater than 20 per cent between males and females would be expected to lead to the exclusion of the larger animals from swards where grazing pressure had been sufficient to reduce vegetation height to below some critical level.

Incidentally, these same allometric relations between different body parameters also provide the explanation for our earlier observation (footnote, page 62) that large animals are more capable of tolerating poorer-quality foods than are small animals: as body size increases, gut capacity increases at the same rate, while actual nutrient requirements, although themselves increasing, do so relatively more slowly, being related to $W^{0.75}$.

SOCIAL ORGANISATION IN RELATION TO HABITAT

Among those species where sociality may be 'tolerated', we have seen that social groups are often impermanent and group composition and size may change with season and circumstance. Sizes of groups formed in any particular instance are in fact strongly influenced by environment, so that there are often marked differences in the size of social groups to be found in different areas. In particular, differences in group size can be seen to be related to those same major aspects of resource use that we considered in the last chapter: habitat and diet. Many of the differences in social organisation *between* species can also be explained in relation to differences in habitat use; and this may be another reason why musk deer and muntjac and their relatives have remained essentially solitary species, while it is among the Odocoilinae and Cervinae that more social complexity has evolved. Muntjac and musk deer are creatures of dense forest, and solitary nature and small size are characteristic of such environments. Even among the more advanced deer groups, those species which have specialised to the denser forest habitats (e.g. brocket deer, *Mazama* spp., or pudu) have retained a solitary character. It is only among those species of woodland edge or open country that larger social groups have been formed.

Thus, while there may appear to be an evolutionary trend towards increasing sociality among the deer, such a relationship may well be purely secondary. We described in Chapter 2 that evolution within the Cervidae was accompanied by a movement away from the dense forests of their origins, so that, while the more primitive forms are still restricted to the forest environment, at least some of the more advanced species have adapted to more open habitat such as woodland edge or even, in the extreme, open grasslands. A shift from forest-dwelling to living in more open country would in itself be accompanied by an increase in the size of social groups, effected by that change of habitat alone. Since that change in habitat use is associated with evolutionary development, the accompanying trend to increased group size and apparent sociality would also appear to be an evolutionary progression, although in practice the primary relationship is between social organisation and habitat.

Group size and 'character' of the social system adopted by different species have been shown in a number of animal species to be closely related to environmental circumstance, to be precisely adapted to ecology. To understand this we must appreciate what living in a group entails. Most animals are essentially individualistic: to come into close contact with others of their own species in any form of social group exposes the individual to a number of potential disadvantages — competition for food, for mates, or other resources, potential interference. Other animals around it may not only displace it from favoured feeding areas or deny it access to other resources, they may also get in its line of sight, or make such a noise that it is not so well able to detect predators or other potential danger.

On the other hand, there are a number of potential advantages in being part of a group. With more eyes and ears alert for predators, the chances are that potential enemies will be detected far earlier than by the individual alone; the individual is in any case 'diluted' by being one of a larger group, and by the pure laws of chance it is less likely to be the one that is selected by a predator. Finally, groups can often offer more positive defence against predation through joining together into a counter-attack, taking advantage of the strength offered by numbers, or by showing some defensive behaviour feasible only for groups rather than individuals (as, for example, buffalo or musk-oxen group together in a circle, with cows and calves in the centre, and the bulls in a tight ring around them with horns swinging outward).

Living as part of a group can also increase foraging ability. As one consequence of the greater number of eyes and ears alert for danger, each individual needs to spend less time alert itself and can spend longer feeding. Further, there are a number of situations in which groups can exploit more efficiently particular types of food or particular food distributions than can individuals. Where food is very sparsely and patchily distributed in the environment (but each patch is rich), an individual may spend hours foraging without encountering a food patch at all; if and when it does, there may be superabundant food. If a group were to spread out more evenly over the ground, the chances of one group member encountering a food patch are much increased: so long as the food patches, once located, are rich enough to provide for the whole group, then grouping will prove advantageous. Furthermore, it is not just distribution of food that may matter: co-operative foraging in groups may enable the animals to exploit particular types of food that they could not exploit on their own (as pelicans may surround a shoal of surface-feeding fish and form a living 'net' to prevent the fish escaping, while all feed on the captive shoal). Finally, living in a group may provide reproductive advantages, such as regular access to the opposite sex. (For a more detailed review of all the advantages and disadvantages of group-living, see Clutton-Brock and Harvey, 1978; Putman and Wratten, 1984.)

The actual social system adopted by any animal species — that of being solitary or social — will depend on the relative importance of each of the different merits and demerits when applied to its own particular circumstances. Among those species which opt for group-living, the size of group adopted will be determined in relation to the same essential considerations, to be such as to optimise advantages while minimising disadvantages, and clearly the balance which results will be different in different situations. The anti-predator advantages of living in a large group may outweigh disadvantages to the individual in open country with little cover, where an early-warning system and group defence may be the only effective strategy. In dense forest, however, a group is very obvious and is, in any case, hard to maintain: perhaps the best strategy then is to escape detection in the first place — to be secretive, elusive, and solitary. Likewise, where animals feed on small discrete food items relatively abundant in a small area, and thus where each individual can be supported by a relatively small home range, perhaps the best option to ensure continued free access to food is to remain solitary within that area, and to defend it against all other conspecifics. The smaller the area, the easier it is to defend, increasing the selection pressure towards solitary habit. In environments where food is more patchily distributed, as we have seen, larger foraging parties with larger ranges may be of advantage. Where animals tend to feed relatively unselectively and food is thus abundant and readily available, food itself may no longer be the factor of prime importance in determining group size, and predation or reproductive considerations may assume greater import. Overall, it is clear that, to satisfy nutritional, anti-predator and social requirements from the particular 'facilities' on offer, group size becomes precisely adapted to habitat type.

Perhaps the best-known illustration of this relationship comes from the studies of Jarman (1974) on East African antelope and gazelle. Jarman showed quite clearly that, for any species assuming a particular foraging style within a particular habitat, there was a particular group size to be adopted; almost irrespective of the species concerned, a particular group size appears characteristic of a given life style (see Figure 10.6 in Putman and Wratten, 1984). Thus browsers in dense forest should be solitary or run only in mated pairs. As concentrate-selectors (pages 44–8), they are highly selective in what they feed on, in terms of types of food or parts of plants, but relatively unselective with respect to species. Such needs can be met in a small home range, but to restrict competition from other conspecifics that range should be defended as an exclusive territory. Within dense forests, it is impracticable to try to maintain the cohesiveness of a larger group and thus anti-predator advantages of a large group cannot be realised; further, with dense cover, the optimum anti-predator strategy may in any case be to be secretive and solitary, to rely upon escaping detection rather than adopt a strategy of active defence. By contrast, out on the open grasslands, relatively unselective bulk feeders are offered an abundance of food; group size will not greatly influence feeding efficiency either positively or negatively. Such grazers are, however, exposed and far from cover. The best anti-predator strategy is thus to opt for all the advantages of early detection and group defence offered by being a member of a larger social unit. Intermediate examples can be explained in the same way, and as a general rule we may conclude that group sizes increase as the environment becomes progressively more open (Estes, 1974; Jarman, 1974).

Differences between the social systems of different deer species can be accounted for on exactly the same basis (Table 4.1). As we have noted, species such as the brocket deer or pudu of South American rainforests are small concentrate-selectors of dense forest, and tend to be solitary. It is among species such as the pampas deer, the barasingha or the rusa deer of Asia, which have become adapted to open grasslands, that we find the largest social aggregations. Species of open woodland and woodland edge (like many of the *Cervus* species, the fallow deer and the chital), when they group at all, tend to be found in groups of intermediate size. This close relationship apparent between group size and both habitat and feeding style is extremely important, for it offers a fundamental explanation for why some species of deer are solitary while others adopt a more social habit.

The relationship between group size and environment does not, however, result just in differences in social grouping between different taxonomic species. Among those species, such as roe deer, hog deer, rusa and chital, which are more flexible in their habitat requirements and can be found in a wide variety of rather different community types (pages 37–8), the same great variation in group size may be observed: in this case between different populations within one single species, as each local population adopts the social structure appropriate to the habitat in which it occurs.

One of the most extreme examples of how social systems may adapt to environmental circumstance, and indeed of how flexible individual species may be in terms of social organisation, is offered by the European roe deer. A relatively primitive member of the Odocoilinae, we might expect it to be a solitary, asocial beast, a conclusion we would also draw, independently, from the observation that roe are true concentrate-selectors, characteristic in the main of dense woodlands with deep cover. Such an expectation is realised in practice, for when roe are studied in such environments they are relatively solitary, and are strongly territorial (e.g. Andersen, 1961; Cumming, 1966; Strandgaard, 1972; Hosey, 1974; Ellenberg, 1978). Roe deer, however, are opportunists, as we have already discovered. While their feeding style confines them to relatively early successional environments, their small size and relatively modest food requirements in terms of actual bulk allow them to exploit a variety of potential niches. They may be found in small coppice

Table 4.1 Group size of different species of deer related to ecology. Sources: various (see text)

	Feeding style	Habitat				
		Dense forest/ thicket	Open woodland	Forest edge/shrub	Open grassland	Marshland/ carr
Concentrate-selector	Fruit-feeder Selective browser	Mazama: solitary/ pairs Pudu: solitary Muntjac: solitary/ pairs Huemul: solitary/ pairs Roe: solitary/ small groups		Huemul: <5	Roe: up to 40	Chinese water deer: 1–2 Moose: 1–2
	Browser/ grazer	Sambar: solitary/ small groups Hog deer: solitary/ small groups	Sambar: 3–5 White-tailed deer: 1–3		Hog deer: 10–20 White-tailed deer: 6–10	
Intermediate feeder	Grazer/ browser	Red deer: 1–3 Sika: 1–3	Fallow: 1–10	Chital: 5–10	Chital: 15–20 occasionally 100s Red deer: 1–50 Fallow: 30–100	
Bulk feeder	Grazer				Pampas deer: 5–15 Barasingha: 12–100s Rusa: 100s	Eld's deer: 'large herds'

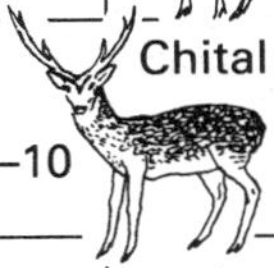

woodlands in agricultural areas, where they remain solitary while in the woodland itself, but may form temporary associations of up to seven to ten animals when foraging out in agricultural crops. As described on page 38, roe have also managed to adapt to the totally agricultural landscapes of Middle Europe. Here, in perfect vindication of our theories so far, they may gather over winter in permanent social groups of up to 60–70 individuals (Bresinski, 1982; Kaluzinski, 1982). Similar responses are seen in other species (Table 4.1). Among white-tailed deer, group sizes are found to be far larger in populations characteristic of more open habitat than in areas where their ranges are predominantly within woodland (Hirth, 1973; Lagory, 1986). Hog deer, while characteristically associated with grass jungles by the banks of rivers, where they are seen in groups of between 10 and 20 individuals, are also found in denser scrub and even within forests (e.g. Mishra, 1982), when they occur singly, in pairs or in family groups of three to five (Kurt, 1978). Chital of forest edge and wooded grassland are typically recorded in groups of between five and ten individuals (with a mean group size of 7.5: Mishra, 1982). In open savanna, however, they may attain groups of tremendous size: Kurt (1978) quotes observations of chital groups in such areas of 800 or more individuals, but these are somewhat extreme and more probably reflect only temporary feeding aggregations (page 59). Nonetheless, the fact that these aggregations are permitted to form at all is clearly a function of habitat, for they are not recorded in more 'closed' environments.

A similar variation in group size is found among fallow deer in Britain (e.g. Putman, 1981; Waterfield, 1986) and elsewhere in Europe (Schaal, 1982). Once again, groups within the more dense habitats (thickets and dense coniferous plantations; young, unthinned blocks of deciduous woodland) tend to be small (two or three individuals), with groups in open-based woodland (mature conifers or broadleaves) or in glades and clearings tending to be larger (up to seven or eight animals). When the deer leave the woodland to forage in fields or other open ground, separate social groups may coalesce to form aggregations often composed of 30–40 animals and on occasions numbering up to 150–200 individuals in favoured feeding grounds. Here, however, habitat is affecting more than just group size. Schaal (1982) reports from Alsace that in separate populations of fallow deer, some living permanently in woodland areas, and others resident in areas with relatively less available woodland but correspondingly a greater proportion of their range in rather open ground, not only did the size of groups change (in the now familiar direction of an increase in size as the envrionment became more open) but the structure of those groups also altered. Fallow deer are one of the species in which, typically, the sexes remain separate for much of the year, meeting only to mate (page 61). Normally, therefore, social groups are composed entirely of males, or of females with their offspring (such groups may include males up to the age of 18 months or so). Only when large feeding aggregations form when groups of animals come together on favoured feeding grounds, are groups containing adults of both sexes encountered. These groups are, however, impermanent, and such aggregations do not represent a true social unit but rather a loose association of a number of separate smaller groups, each retaining their separate identity within the whole, joining and leaving the aggregation as a unit; and each of these subgroups retains the typical separation of males and females. In Schaal's studies in Alsace, however, it became clear that, not only were group sizes in the more open environments larger than those of more closed habitats, but that groups containing adults of both sexes were encountered fairly regularly throughout the year and not solely within the context of the larger feeding aggregations. These observations have since been confirmed and expanded (e.g. Waterfield, 1986; Thirgood, in prep.) and it is clear that, for fallow deer at least, the type of environment in which they find themselves affects not only overall group size, but also the size of the individual social units which go to make up any larger aggregation that that may represent, and in addition the structure of those sub-units, with groups containing adults of both sexes encountered far more regularly in open environments than in their 'traditional' woodland habitat.

The precision with which group size is matched to habitat, and the flexibility of that adaptation, are further illustrated at a level of resolution one stage higher still. Among those species whose environment is not strictly homogeneous, but offers a mosaic of different vegetation types patchworked together, group size does not even remain constant within any one population, but may change from day to day, even from minute to minute, with changing patterns of habitat usage — reflecting at all times the vegetation type occupied at that particular juncture. If permanently resident in one habitat type, groups will be of constant size appropriate to that habitat, producing the responses we have just described. In mixed environments, however, where animals may exploit a range of different vegetational types, group size will actually change as they move from one habitat to another. In the example we have just cited, fallow deer which live permanently in more open environments tend to occur in larger groups than those whose range is predominantly within woodland; but, where animals move between woodlands and open fields, groups join with each other and split again as they move from one habitat type to another.

Red deer in woodlands are commonly encountered individually or in pairs; in mixed environments, where they may move between a variety of different habitats, groups form and break. When in dense woodlands, the deer still operate essentially as individuals, but they group together into larger associations in clearings, glades or more open woodland, and into larger groups still when they forage out onto open moorland. Individuals join together as they come into the more open areas; groups fragment when the deer move back into denser woodland. It is the habitat at the time that determines the tendency to group, and the size of the group formed and social structure are extremely flexible.

Such flexibility is not, however, universal. In the north of Scotland the same species occurs the year around on open moorland. Individuals come together into sizeable groups, but, since the animals do not move between this open habitat and others, the larger group size remains continuously appropriate. This may be the explanation why the Scottish red deer studied by Clutton-Brock and his co-workers on the Isle of Rhum were found to form relatively permanent and persistent social units, with clearly persistent matrilineal groups as the fundamental unit: a shift in this case not only in group size, but in social organisation, from temporary association to formation of relatively more permanent social groups.

In fact, it would appear that environmental circumstance may not only result in a shift from solitary to social as we have just described, but may also affect type of sociality among red deer. Where the deer are in woodland all the time, they act as individuals. In complex environments with varied habitats, groups may form in open areas but fragment when the deer are exploiting woodland; no unit larger than the individual will persist for long in either case. In the unusual situation provided by the uniformly open range of Rhum, larger groups will be continually appropriate. Under these circumstances these groups may then become more permanent as actual units, encouraging the development of group-related behaviours such as matrilineal dominance hierarchies, etc., while woodland reds or reds of mixed environments remain more individualistic. Kurt (1978) found similar shifts within species of Asian deer between his categories of 'solitary but aggregational' and 'aggregational hierarchical' (page 61) correlated with habitat changes (Figure 4.1).

Thus, in the same way that a shift from closed to open environments may be responsible for a change from solitary to aggregational habit

Figure 4.1 Adaptation of social organisation of south Asian deer to habitat condition and population density (modified from Kurt, 1978)

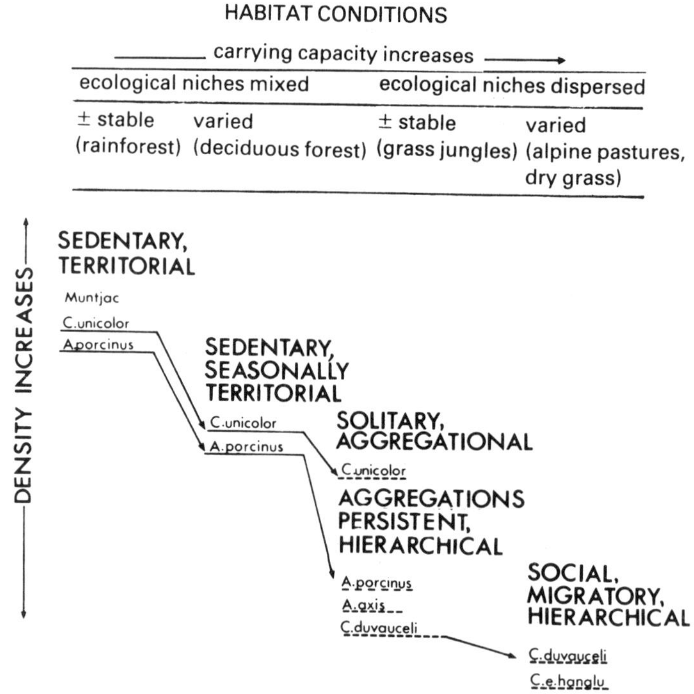

(Table 4.1 and pages 69–9), differences between the more social species in the permanence of association and thus degree of sociality may also reflect environmental circumstance: but in this case relating more to homogeneity or heterogeneity of the vegetatational mix in which the particular species or population is found. Temporary aggregations within which each animal remains essentially individualistic will be characteristic of environments offering a complex mosaic of open and closed community types; more truly 'social' groupings, within which may be observed true hierarchies, will be developed in environments both open and homogeneous.

Differences in Group Sizes of Males and Females

One final little subtlety of the effect of habitat and habitat use on social structure develops from the tendency among deer towards segregation between the two sexes. Since, among the more social species, this is expressed through the establishment within a population of distinct male 'bachelor' groups and groups containing only females and followers, structuring of these groups — in their size and composition — is of course effected within the group (and thus within the sex). Sizes of male and female groups in given seasons may thus differ as males and females respond to the specific pressures of their environment (see for example Figures 5.1 and 5.2). If males and females occupy different habitats, this will effectively mean that their group sizes, adapting to the environment in which the animals themselves occur, will differ. Even within the same habitat, if males and females adopt different feeding styles (page 50; Staines *et al.*, 1982; Clutton-Brock *et al.*, 1982) this, too, may lead to subtle differences in group sizes appropriate to the two sexes.

Group Size and Season

Group size within a species is markedly influenced by habitat, and populations characteristic of different environments show clear and predictable differences in the size and structure of social groups. Group size, however, also shows seasonal change. To some extent this is a further consequence of the influence of habitat, for many of these seasonal changes in social organisation can be explained by seasonal changes in the pattern of habitat use (Chapter 3). Other changes occur independently of this and relate primarily to the breeding cycle. Females of all species tend to become rather solitary just before the birth of their offspring and for the first few weeks of the baby's life, even if they later do join up into the larger groups characteristic of the more social species. Males tend to become solitary and secretive at that time of year when they shed their antlers and prior to their regrowth. Of course, during the period of the rut, when the animals come together to mate, group sizes and structures may alter again as the particular mating groups are formed. These seasonal changes in social organisation will be considered in more detail in the next chapter.

SEXUAL ORGANISATION

Associated with the differences in social organisation which may be recognised between the different species of deer is a complementary change in sexual behaviour.

There is in fact a general evolutionary trend with the Cervidae from a monogamous mating system (whether or not that is accompanied by the development of a true pair-bond) towards an increasing degree of polygyny. The influence of such an evolutionary progression is hard to quantify, however, for it is once again difficult to separate it from the exactly complementary effects of habitat on social organisation that we have just reviewed. Musk deer and muntjac may indeed mate with only one or two females in any season: but is that because they are evolutionary primitive, or because they are forest-dwelling concentrate-selectors which, in consequence, adopt a solitary habit or at most consort in pairs? Whatever the true cause for these particular animals, even among the more advanced Cervinae and Odocoilinae any species of a solitary nature is, because of its habitat and life style, restricted in mating opportunities. If territorial (which would be the norm for such species), it will either form a monogamous pair-bond with one other individual or, more usually (for in general the two sexes tend to sort out their territorial boundaries independently of each other: page 74), will mate in due season with those animals whose ranges its own territory overlaps. In those species whose ecology permits or indeed favours the formation of groups, there is, however, as we have suggested, a greater potential for polygyny, and indeed selection will favour this. As in most mammals, male deer contribute little to the breeding attempt beyond copulation (although in some species males may guard their females against harassment by other males, or against predators). There is thus no real advantage in females 'having a male to themselves'. Accordingly, females should choose to mate with the fittest male available (to ensure the maximum fitness of their own offspring) and, since there is no disadvantage associated with sharing a male, the system will tend towards polygyny, with the 'best' males having multiple partners and weak males failing to mate at all.

Among the more social species, therefore, polygyny is the norm and there is intense competition among males for the right to mate and to attract females. This competition and self-advertisement may take a variety of forms. Males of some species, such as red deer or reindeer just to take two examples from many, fight among themselves during the breeding season for temporary 'ownership' of harems of females. In other species, females themselves are not owned or defended, but access to mates and the chance to advertise are conferred upon a male by ownership of land. In species such as the fallow deer, males compete to set up rutting stands (exclusive display grounds: effectively small temporary breeding territories) to which they may attract females. In some cases, the 'territory' secured may diminish to the stage where it is vanishingly small; males congregate in traditional areas — like the communal leks of ruff, or black grouse and other gamebirds — and compete among themselves merely in trying to out-advertise their rivals in their display to receptive females.

The system adopted varies from species to species, and in some cases it is clear that it is the only mechanism appropriate. Barren-ground caribou, for example, sort out their mating title during the migration from winter to summer feeding grounds. In such a situation, where they are continually on the move, ownership of land would be out of place; maintenance of a harem of females is clearly the best strategy. Similar considerations would apply for species undertaking even relatively local migrations and may explain the adoption of such a strategy in many of the other harem-holding species, too.

Where populations are resident within a relatively small home range, the distinction between possession of a harem and possession of land in the breeding season becomes a fine one, but true rutting stands tend to be really rather small in area, and a stand-holder is more concerned with defence of that breeding site than with defence of females attracted to it. Indeed, in many cases, once a stand-holding fallow buck has mated with females attracted to his display ground, he makes little attempt to retain them within his stand. The females may of their own accord remain within the vicinity, but the buck makes relatively little attempt to herd them or to keep them there and they not infrequently drift off — and may even join the rutting group of another male further on.

Lekking is a curious phenomenon not regularly encountered in any deer species. Where it does occur among ungulates, in general it is found in particular populations of species which would normally hold a rutting stand or territory. Among the deer, lekking has so far been reliably reported only for fallow deer (Schaal, 1987). Although mature fallow bucks typically occupy an exclusive rutting stand, which is defended aggressively against most potential rivals, they quite frequently tolerate the continued presence of one other, particular, male on their 'patch' (often, but not necessarily, a younger buck), and such 'satellite' males are by no means uncommon. Less frequently, multiple stands with up to five or six males displaying have been observed; and, very occasionally indeed, a whole system of stands may be established very close to each other, with as many as 20 or 30 adjacent rutting stands in an area of only a few hundred square metres (Schaal, 1987). In such situations, the area defended decreases to become infinitesimally small, and the deer effectively operate as a true lek. The reasons for this occasional 'aggregation' of rutting stands in fallow deer, or for that matter among other ungulates, are unclear. (Even among those populations which do show this form of mating system, leks do not form every year.) Perhaps females in these areas are at particularly high densities, or suitable habitat for display grounds is scarce in that particular area; whatever the cause, competition for females in these areas must be extremely severe.

HOME RANGE AND TERRITORY

In talking about those factors which may influence group size and social system, we have, in passing, noted a number of other related characters. Thus, in establishing that, as a general rule, forest-dwelling concentrate-selectors tend towards a more solitary habit than do intermediate or bulk feeders, we also suggested that such small, concentrate-selecting species will tend to have small home ranges and to be territorial. Here in effect is another variable related to the whole complex of life style, habitat and group size. Territoriality — defence of a small and exclusive home range and the resources within it — will be favoured by the very same considerations that induce a solitary habit. For we found this solitary habit adaptive as offering the best anti-predator strategy in dense cover, but it was also associated with efficient foraging in such environments. The more of a concentrate-selector our species, the more it is able to sustain itself within a small home range. Yet to do so effectively it must both be solitary and defend the resources represented by its territory. By contrast, animals with less predictable food resources may respond by increasing group size, to have more individuals to search for food patches, but must also cover larger ranges to ensure that they will encounter acceptable food at all times; these species cannot afford to be individually territorial, and usually cover sufficiently large home-range areas that it would be impracticable to defend them even as a 'group territory'. If we look again at Table 4.1 or Figure 4.1, we see that concentrate-selectors of dense forest are indeed not only solitary but territorial, and that the degree of territoriality decreases as the size of home range required to support the particular feeding style increases. Some species are territorial all year round, others, such as roe or sambar, are seasonally territorial; and, by the time we move to species of more open country or more varied habitat, while animals may still remain asocial, and still maintain a defined home range, ranges overlap considerably and are not defended as territories (the 'solitary aggregational' species of Figure 4.1).

In all those species which maintain exclusive territories, territorial boundaries are sorted out between individuals of the same sex. That is, males will carve out their territories with respect to those of other males, while females, which are also territorial, will independently establish their territories in relation to the territorial boundaries of other females. A map showing territorial boundaries of males or females alone would thus show clear separation of all the different individuals, with clearly distinct territories patchworked together, each abutting the other, but showing little or no overlap (Figure 4.2 a and b). Were a map prepared of the territories of males and females in the same area, great overlap between the territories of the two sexes would, however, be apparent (Figure 4.2 c), for each effectively operates independently of the other.

Figure 4.2 actually presents a slightly artificial picture, for it is in practice unusual to find absolutely no overlap between adjacent territories even within one sex. Some overlap does inevitably occur, and although this is usually kept to a minimum there are sometimes exceptions, where unusually high overlap is tolerated without contest. This is more common among females than among males, and where it has been investigated in muntjac, in roe deer and in white-tailed deer (which, I confess, I always consider as the New World's equivalent of the roe) it appears that it is due to mothers permitting their daughters to remain within the parental range when they reach adulthood (Loudon, 1979; Hirth, 1973; Marchinton and Hirth, 1984; Poutsma, 1987).

In environments providing consistent food supplies which are easily located, or where the species concerned are relatively unselective feeders, territoriality is unnecessary. Such species tend to be grassland feeders and, in the open environments of which they are characteristic, anti-predation responses will in any case tend to outweigh any feeding considerations and will result in a further increase in group size; again, territoriality would clearly be inappropriate and ranges are shared. As a result of large groups feeding together, however, food resources can rapidly become exhausted. Here, then, is a further consequence of our feeding style/habitat complex: bulk feeders of open grasslands may tend not only to occur in groups in shared home ranges, but may show regular migrations around their range (again, refer to Figure 4.1). Such a response is, of course, exaggerated in highly seasonal environments with defined and limited growing seasons where regeneration of the depleted food resources may not occur until the following spring or the next rains. In such environments the grazers must then migrate over greater distances still, in literal pursuit of pastures new, and many species of

Figure 4.2 Territory maps of a number of male and female roe deer in a Dorset woodland during 1968. (a) Male territories are non-overlapping and ranges are assorted entirely with regard to the territorial boundaries of other males; (b) females' range boundaries are also drawn up with regard to those of other females, although some overlap may occur between the ranges of related animals (text: page 74); (c) ranges of males and females show complete overlap. Range maps are taken from data of Bramley, 1970, and redrawn here, though not all ranges of all individuals known to be present in the wood at the time are represented; to preserve simplicity in the figure, some known territories have been omitted)

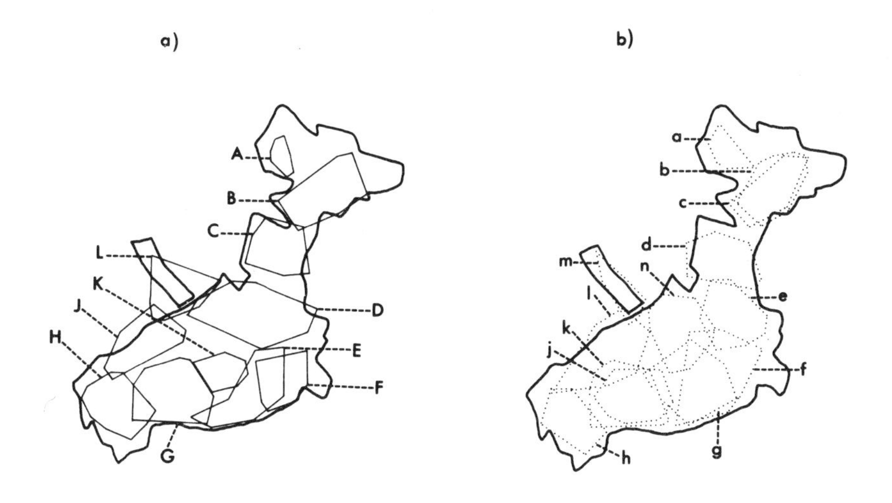

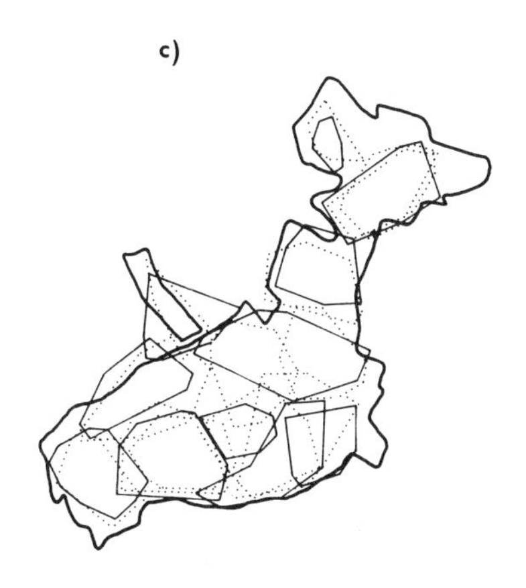

deer undertake 'local' migrations between distinct winter and summer ranges, or wet and dry-season ranges: occupying different areas in response to the quantity and quality of available forage, or indeed its accessibility. This is in practice merely an extension of seasonal changes in habitat use described on pages 38–42, and may of course occur irrespective of group size in such situations, as food in one part of the animals' range becomes seasonally unavailable for reasons other than their own heavy grazing (as when, for example, annual grasses die away to dry stalks of no nutritional value whatsoever in the dry seasons of tropical countries, or forbs and shrubs become buried deep in the snow in temperate and more northerly regions).

Thus the barasingha studied by Claude Martin within the Kanha National Park in central India moved between distinct wet- and dry-season ranges 7 km apart in response to changing availability of food and water (Martin, 1978). Moorland populations of Scottish red deer frequently undertake similar local migrations, moving to lower ground or into woodlands during winter to seek food and to find shelter from adverse weather. Many populations of their American cousins, the wapiti, undertake similar local migrations between winter and summer ranges in North America, as do black-tailed and white-tailed deer (e.g. Hanley, 1984; Jenkins and Wright, 1988).

The most renowned migrations among the deer are of course the long treks undertaken by northern populations of reindeer or caribou from the woodlands of the taiga in which they spend the winter to the lush pastures of the summer tundra. Once again, the southward migration is enforced by the snows which cover the remaining vegetation (although in this case the tremendous energy cost of the heat loss the deer would sustain during the arctic winter clearly contributes an additional factor to the equation).

Range Size

From what we have established thus far, we would expect the ranges/territories of small, forest-dwelling concentrate-selectors to be smaller than the ranges of intermediate feeders, which in turn would be smaller than those of bulk feeders. Such an idea is evident in fact as well as in theory. Clearly, it is valid to compare the ranges only of animals in similar geographic regions (for other factors such as gross rainfall, temperature, etc. might also affect *absolute* range size of all groups), but within such zones we may indeed observe such a series. In the UK, roe deer in woodland habitat have ranges of between about 5 ha and 15 ha; fallow deer range over areas of some 50–250 ha (males) or 50–90 ha (females); and red deer in Scottish forests, feeding more as bulk feeders than the other species, have ranges of between 150 ha and 460 ha (Hinge, 1986), with males again occupying larger ranges in general than females. (Roe deer in the same areas occupy ranges with a median area of 57 ha.)

Just as we have noted in other elements of behaviour and ecology, however, pronounced variation within species dependent on habitat, comparable almost to the variation recorded between species, so once again range size of any given species is very dependent on environmental type. Thus, in our last example, while red deer in forested areas of Scotland may have ranges of 150–460 ha, with a median area of 241 ha (Hinge, 1986), populations of open moorland and heath may have ranges as large as 600 ha (Staines, 1970; 1974). Similarly, home ranges recorded for different populations of white-tailed deer in the US (summarised by Marchinton and Hirth, 1984) varied between 59 ha and 520 ha in different areas; and roe deer territories have been recorded which were as small as 4 ha (Loudon, 1979) or in other areas as large as 60 ha (Figure 4.3).

Much of this variation may be found to relate to the 'quality' of the habitat in terms of its ability both to provide a sufficient quantity of nutritious forage and to provide shelter. Johnson (1984) showed for roe deer in England that range size was related primarily to food resources. Range sizes of animals in one of the populations he studied were significantly larger than (nearly double) those of animals from a second site. Estimates of food resources contained within animals' territories in the two sites suggested that those at the first site (where ranges were larger) contained a greater biomass, and indeed greater digestible biomass of potential food; the foods were, however, of different nutritional quality, and, when analysis was confined to the availability of digestible nitrogen (see page 48), adult ranges in the two study areas were found to be equivalent. Range sizes of red deer on the Isle of Rhum in Scotland were likewise shown to be correlated with the relative proportion of heathland and grassland contained, so that range size increased as the proportional area of (preferred) grasslands declined (Clutton-Brock *et al.*, 1982).

Cover may also play a role. Sticking with the same two species: roe deer in agricultural areas of Europe are presented year round with abundant supplies of highly digestible forage, yet ranges must be large enough to include an area of cover for resting and rumination. Red deer on Rhum occupy far smaller ranges than those of equivalent hill range on the mainland of Scotland (Staines, 1970; 1974) and this, too, is believed to be due to the fact that Rhum provides a finer-scale mosaic of different habitats, with many small feeding and sheltering areas patchworked together in a smaller total area; in such a fine-grained environment, so long as foraging needs are met, home range need not be further increased to ensure inclusion of suitable sheltering areas, while in the coarser-grained mosaic of the mainland larger areas must be covered in order to gain access to shelter (Lowe, 1966; Mitchell *et al* ., 1977).

Not only is there variation within species in relation to environment, there is often in addition a pronounced difference in the range sizes of males and females. Generally, males occupy larger territories/home ranges than females (page 76 for fallow deer; Figure 4.3 for roe) and this difference may be in the region of a two-fold increase in size. Reasons for the difference in range size are complex. Although we have noted among the deer a clear size dimorphism (with presumably a corresponding difference in food requirement), this is insufficient on its own to account for the magnitude of difference observed in home-range area. Further, despite the difference in size, we might in fact expect females to have somewhat higher energy requirements than males, since through most of the year they are incurring the extra demands of pregnancy and lactation (and energy requirements may rise by up to 1.5 times those of non-breeding females: Weiner, 1977). It seems unlikely therefore that the larger size of male ranges is related in any simple way to nutrition; it is more likely a function of social needs — as for example in providing access

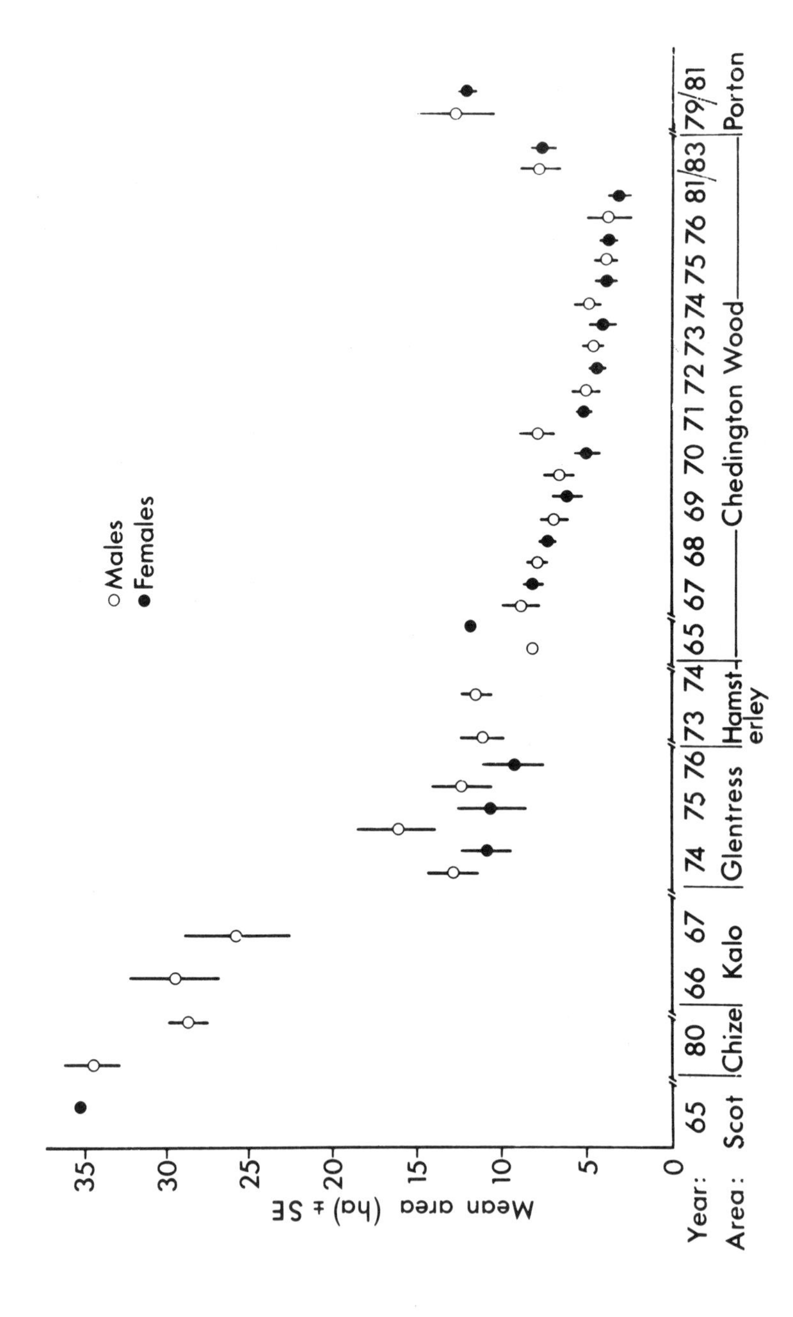

Figure 4.3 Variation in range size of the European roe deer in different environments (data summarised by Johnson, 1984)

to breeding females. Where territories are defended, the larger size of male territories may also be an adaptation to prevent other males interfering in the territory-holder's access to females, while at the same time increasing the number of females included within an individual male's range. Increasing the size of a defended territory not only increases the number of females included within it, it also effectively reduces the number of other males able to obtain territories (and thus to obtain access to mates). As a result, by holding a large territory a male not only increases the number of offspring he leaves to succeeding generations, he also reduces the breeding potential of other males; by both actions he increases the proportion of his contribution to the next generation of the population, and Verner (1978) has suggested that this 'dog-in-the-manger' attitude, this suppression of the breeding potential of other males, may in itself be an important reason for the larger range size of males in territorial species. In non-territorial species of course such arguments will not apply, since home ranges show extensive overlap and individuals freely tolerate the presence of others. Here, too, however, larger range size in males would offer them access to larger numbers of potential mates, and it seems probable that such social factors underlie the greater range sizes of males in all species.

Figures quoted above for actual range sizes are presented as the total area which may be used by individual animals. Commonly, however, the deer restrict most of their day-to-day activity to a much smaller area. If we consider as a 'core range' that area within which the animal actually spends 80 per cent of its time, we find that it is much reduced. Core ranges of fallow deer in the New Forest in England have a mean value of 63 ha (males), while the equivalent mean value for the total range is 153 ha (Rand in Putman, 1986). For roe deer in Johnson's (1984) studies, core areas were again some 44 per cent of the total range calculated.

Even within these core ranges, animals rarely use the whole area at any one time. A larger foraging area may be needed in winter (or corresponding dry season) than during the full growth of spring and summer or rainy seasons, so that the range may expand and contract in area as the seasons pass. Alternatively, the animal may use quite different parts of its overall annual range at different seasons, having distinct and quite separate wet- and dry-season ranges, let us say (as the barasingha in Martin's (1978) study in central India). In either case, whether the ranges used at different seasons are inclusive or exclusive of each other, actual range size changes with season, and the range size of a given season may be far less than suggested by even calculated core areas of an annual range. New Forest fallow deer have far smaller ranges in summer than in winter: while winter ranges for males have been quoted with a mean area overall of 153 ha and a core of 63 ha, ranges in summer have a mean of 107 ha and a core of only 35 ha. Martin's barasingha have distinct, non-overlapping wet- and dry-season ranges each of perhaps 100 ha, but a total annual range of nearer 46 km^2.

(Even within a seasonal range each animal will have preferred areas for foraging and preferred areas of shelter, and uses the resources offered by its home range in a relatively predictable way from day to day. There may be some variation, with different foraging areas used perhaps under different weather conditions, depending on prevailing wind direction or whether or not it is raining, but in general the daily range is not that much smaller than the overall seasonal range. Gent (1983), radio-tracking roe deer in a woodland environment in Dorset, found that in any 24-hour period individuals used up to 40 per cent by area of the total seasonal range.)

Seasonal change in range use may also be reflected in the degree of overlap observed between ranges of individual animals. While we have so far in these pages considered the roe deer as a territorial species, its territoriality is far more pronounced in summer (April–August) than in winter. Food is of course likely to be in shortest supply over the winter period and individual animals may have to range far more widely to procure sufficient food. Under such conditions maintenance of territorial boundaries becomes difficult, as the area is too great for one animal to defend adequately — and in any case perhaps becomes inappropriate. In practice, however, roe deer do not seem to show a great increase in range size in winter, with winter ranges not significantly different from those of summer (e.g. Johnson, 1984) or even smaller (e.g. Gent, 1983), so the breakdown of territoriality may not in fact be related to a need to range further for food in winter. Since roe deer mate in early summer (April–May), when territoriality is at its most intense, it is perhaps possible that territoriality in this species is related more to breeding requirements than to foraging (and see also p. 79). At all events, whether in woodland or in the agricultural landscape of Middle Europe, roe show greater tolerance, less territorial aggressiveness, in the months of winter.

Social Behaviour

Whether solitary or gregarious, individuals interact with others. Territories must be contested, marked and defended, members of a social group must be recognised and, where appropriate, group cohesiveness maintained. Males must compete with others to secure access to mates, and must attract those same mates. Deer are not very vocal animals in the main (doubtless because they are so vulnerable to predation), and most communication is by sight, scent, or by direct physical interaction.

Territorial species for the most part advertise territory-ownership by scent-marking. As we noted in Chapter 2, deer have scent glands on the face beside the eye, and also between the cleaves of the foot. Trails and runways through the territory are thus scented during normal use by the secretions of these interdigital glands; additional marks may be added by deliberately scraping at the ground in certain areas or rubbing the facial glands on twigs or other vegetation. Urine may also be used as a scent-marker. Such marks are generally distributed throughout the territory and are not restricted to boundaries; an intruder may thus encounter scents of territory-ownership at any point within the range. Very similar methods are used for defining a 'breeding territory' by males of those species which do not defend a territory all year around, but establish a rutting stand in the breeding season to which to attract females (defending a parcel of land for the females that may be attracted to it, rather than contending directly for the females themselves). Such rutting males also mark their 'territory' by scraping the ground with the forefeet — frequently urinating directly into the scrape — and by rubbing the face on vegetation. Marking with scent from the facial gland may be quite specialised behaviour. In many woodland species, males inflict

1 *Himalayan musk deer*

2 *Chinese water deer*

3 *Male Chinese muntjac*

4 *Red deer stag*

5 *Sika hind in summer coat*

6 *Japanese sika stags*

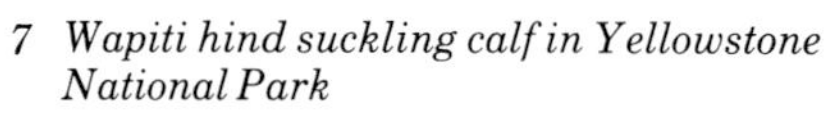

7 Wapiti hind suckling calf in Yellowstone National Park

8 Père David stags in Woburn Park

9 New Forest Fallow bucks grazing

10 A mixed group of fallow deer feeding out in a woodland clearing. Note the simple spike antlers of the young buck on the left

11 Chital deer in Southern India

12 *Hog deer in typical 'grass-jungle' habitat*

13 *Young European roebuck – showing clearly the pronounced 'moustache'*

14 *White-tailed buck*

15 Mule deer doe

16 South American marsh deer

17 Red brocket deer

18 Bull moose feeding in wet forest

19 Reindeer grazing lichens and reindeer moss, South Georgia

20 Reindeer bulls fighting; note how the antlers interlock to offer purchase

21 Bark-stripping damage: a group of sweet chestnut trees stripped by fallow deer, showing the extent of damage which may be caused during hard winters

22 Wild cherry with bark rubbed from the stem by a 'fraying' roebuck

23 Fallow fawn lying concealed in vegetation

24 Fallow bucks fighting during the rut

wounds on mature trees within their stand by scoring them with their antlers. These fresh wounds on the tree trunk may then be anointed with scent from the facial gland. Antlers may be used in territory-marking in other ways, too (both by those species which hold 'territories' only in the rut, and by more strictly territorial species), by thrashing at ground vegetation, shrubs and saplings, thereby uprooting the ground mat or 'fraying' the young trees and stripping them completely of bark and foliage (Plate 22).

While such activities may mark a territory and advertise ownership, and thus may deter intruders, the territory must be taken in the first place and persistent intruders who 'ignore' the ownership markers must be challenged. Territorial disputes among the deer usually involve direct confrontation, with actual fighting commonplace. Those species which are strictly territorial the year around, or as roe or hog deer for a major portion of the year, tend to be relatively small species of dense cover. Signalling displays are likely to be of little value, because visibility is poor, and most disputes are settled by fights — by furious chases or by use of weapons. Musk deer, water deer and muntjac all have, and use, enlarged upper-canine tusks as weapons. Among the antlered deer, territorial species such as the roe, the brockets and again the muntjacs tend to have relatively simple, pointed antlers more likely to be of use in actual attack than in mere display. (By contrast, the antlers borne by males of social species tend to be much larger, curved, and often tremendously branched or palmated, and are clearly more adapted to wrestling matches and stylised trials of strength, or to pure display, than to serious contest: Chapter 7.)

Territorial ownership confers right of access to food, water, mates and all other necessary resources which go with that parcel of land. Among non-territorial species, right of access to essential resources cannot be so simply defined by ownership of some parcel of land to which they attach. Rights to resources, where they are limited in supply, must be resolved by some other method and are most simply determined by establishment of some system of social rank or dominance. Clear dominance hierarchies may be observed in social species of deer, with rank position being related primarily to size (Franklin *et al.*, 1975; Franklin and Lieb, 1979; Suttie, 1979; Clutton-Brock *et al.*, 1982). There is also a clear association with age (Clutton-Brock *et al.*, 1982), but since body size itself also increases with age, at least initially, the effects of age alone are apparent only among mature animals. Rank order (like territory-ownership) is established separately within the separate male and female groups of segregated populations, but in general, when the sexes meet, males have a higher rank than females of equivalent age (a fact indubitably explained partly by the size dimorphism so apparent in the more social deer).

While hierarchies are at their most apparent among those species where social groupings may remain constant in size and composition for quite some period, it is clear that they also exist, even though they may be less obvious, among those species whose groupings are more fluid. Because even for these species the potential 'pool' of individuals with which one may form a group is limited to those which share a geographical range, they will encounter all other individuals with which they might group at fairly frequent intervals and thus will have plenty of opportunity to establish rank. Indeed, even in seasonally territorial species such as the roe deer or hog deer, dominance hierarchies are apparent in those seasons of the year when the animals are non-territorial (Geiger and Kramer, 1974; Poutsma, 1987), although in this case dominance appears to be related to distance of the individual from its normal summer territory.

Ranks are established through threat display and fighting, but fights are more 'conventionalised' than those among territorial species, and involve only limited use of weapons. Instead, disputes tend to be resolved by aggressive display or ritualised trials of strength. Females may threaten each other with bared teeth and flattened ears (Figure 4.4), may bite each other and occasionally rear on hindlegs to box out at each other with the forelimbs. Antlered males — and female reindeer, which also bear antlers — will paw the ground at each other and shake their heads to display the size and complexity of their antlers; they may strut in a 'parallel walk', pacing side by side with a curiously stiff gait and with heads held erect and turned slowly from side to side, again displaying to each other the size of their antlers. Frequently such display is sufficient to resolve the dispute, but in other cases the animals may turn, to interlock antlers head on and engage in protracted wrestling matches.

Actual fighting is usually observed, in either sex, only during the initial resolution of the hierarchy, or when individuals wish to challenge their position in relation to others (much as humans on a tennis-ladder!). Once the hierarchy is established, it is usually adequately reinforced purely by occasional threat. Frequency of contests among males,

Figure 4.4 Threat/dominance display between two red deer hinds

however, increases dramatically during the rut, when individuals are not merely competing among themselves purely for social rank in itself, but are actually competing for a specific immediate resource: access to oestrus females through ownership of a rutting stand or harem. While fighting then becomes common (among red deer on the Isle of Rhum, continuous observation suggested that most mature stags fought at least once every five days during the rut: Clutton-Brock *et al.*, 1979), contests still retain their 'ritualised' form and display represents an important component, far more so than among territorial species.

Describing contests between the Scottish red deer stags of their study, Clutton-Brock *et al.* relate the typical course of events to be that a challenging stag approached to within 200–300m of a harem-holder and the two roared at each other for several minutes, after which the intruder usually withdrew. (Although we noted earlier that vocalisations in deer are uncommon, such roaring, bellowing or whistling by males during the rut is one notable exception. It serves as a threat display or challenge — whose effectiveness is apparent in that the intruders usually do withdraw! — but also as an advertisement to attract females to the harem or rutting stand of dominant males: see Chapter 5. Dominance status appears to be accurately reflected in the nature of the call, and other individuals can assess the dominance of the caller on that basis alone: Clutton-Brock *et al.*, 1979). If roaring displays proved insufficient to resolve a dispute, or if the approaching stag came within 100 m of his opponent, Clutton-Brock *et al.* found that the contest was likely to escalate first to a further exchange of roars, and then, in the majority of cases, to a parallel walk of the form already described. At any moment during the parallel walk either stag might invite contact by turning to face his opponent and lowering his antlers. Opponents almost always accepted this invitation, turned quickly and locked antlers. Both animals would then push vigorously and attempt to twist the opponent off balance to gain the advantage of any slope or irregularities of the ground. Fights lasted until one of the pair was pushed rapidly backwards, broke contact and ran off (Clutton-Brock *et al.*, 1982).

Such fights are, as we have described, clearly conventionalised in form, and many disputes are resolved purely through display. With much male display involving the showing off of the potential weapons represented in their antlers rather than their use as weapons in actual fact, and with such contests as do escalate resolved in the main by wrestling matches head to head with antlers firmly interlocked, the antlers themselves have changed in function, and consequently have altered dramatically in form from the simple spiked weapons of territorial species. The antlers have become extremely large and ornate, often extensively branched, and in some groups (moose, fallow deer) broadly palmated, making their impact in display ever greater. They have become curved into wide, sweeping structures, and this curvature, and the particular arrangement of the branches (or 'tines'), adapt them superbly for being locked together without causing severe injury. Even the mechanical properties (cross-sectional shape of the main beam, degree of curvature, breadth and thickness of load-bearing surfaces) have adapted to the twisting type of stress imposed on them during wrestling bouts. It is the change in function of the antler from weapon to display organ and machinery purely for a particular ritualised type of fight that has resulted in this change in form. The change in function itself relates to the shift from territorial defence in solitary species to social striving in more gregarious species. Since the most extreme contests for dominance are those between males for access to females, and since the severity of male competition increases with the degree of polygyny experienced in the mating system, development of large and complex antler structure can be seen to be related to degree of polygyny in the mating system (Clutton-Brock, 1982). This will be considered in more detail in Chapter 7.

Social interactions between individuals are of course not limited to aggressive or competitive contests for rank or resources. Mothers must recognise their current offspring and develop behaviours to establish a mother-infant bond. Animals living in groups must extend such behaviour to recognition of other individuals of their social clan and behaviours to reinforce the cohesiveness of the group. Recognition of other individuals appears, in deer, to be primarily by scent. Individuals meeting tend to sniff over each other's neck and flanks. As noted in Chapter 2, certain species possess additional scent glands associated with the canon bone of the rear limb. Not all deer possess this additional gland; it appears to be present primarily among the more social species and its presence may well reflect the need for more precise social recognition of other group members in these species.

While we have noted that deer are not particularly vocal animals (except for the bellowing calls of the males during the rut), mothers do maintain contact with their infants through gentle mutters and whickers. In many species, females leave their newborn young lying in protected 'forms' in deep cover when they themselves move out into more open country to forage. Calls to the offspring and its replies to those calls enable the mother to locate her calf on her return, or serve to call the infant to her if she is already in the near vicinity. Such contact calls are extended in the more social species to grunts and whickers within the social group (though these are a phenomenon of female rather than male social groups; outside the rutting period males usually remain totally silent).

The bond between mother and infant is also regularly reinforced by contact, and the mother licks or grooms her calf. These behaviours, too, may be extended within a large social group to maintain group cohesiveness. Once more, such behaviours are more common among female than among male social groups, but in practice mutual grooming and other such contact behaviours are in any case far from common. They may be retained within the constant social units of those species which form clear matrilineal groups (e.g. red deer) but, even here, the greatest amount of grooming is still found between mothers and their offspring, although in this case extended to previous offspring as well as the current calf. In those species apparently no less social, but with less constancy in composition of the groups formed (e.g. fallow deer, chital), such 'group-cohesive' behaviours are infrequent. Groups do not persist for long, individual associations are not strong, and thus the tendency towards such mutualistic behaviours is weaker.

Antipredator Behaviour

One of the major presumed advantages of grouping in open environments is protection from predation (page 64). This advantage is realised by social deer most particularly through advance warning of predators, and in all species, except the most strictly solitary, distinct behaviours have been developed to communicate such warning. In all species harsh alarm barks are recorded. Such barks are, once again, often given by mothers in warning to their offspring of potential danger, but they are not restricted to such context: adults of all species and of both sexes will give these harsh calls of alarm when startled or confronted with sudden danger. Responses vary, with solitary species dashing off into deep cover, and species of open woodlands or woodland edge often scattering, before again running back into cover. In the more open-country species such as pampas deer or reindeer, however, the animals tend to bunch together on hearing the alarm, concentrating in a tight herd while they turn to face the danger. Similar bunching is seen among chital in open environments (Santiapillai, pers. comm).

Alarm barks of this kind are usually accompanied by a series of visual signals. Startled deer often bound away from danger, with the first few steps taken in a curious stiff-legged 'pronking' gait, bouncing jerkily along with all four feet leaving the ground together before relaxing into a more normal step. At the same time the tail may be raised high, and in most species the hair of the caudal patch and rump can be erected. Since the caudal patch in almost all species of deer is of a much lighter colour than the rest of the upper body (and it is often white or very light cream), and erection of this hair increases greatly the apparent size of this rump patch, it is tremendously obvious and acts as a clear signal to all other individuals in the area.

While pronking and tail-flaring are undoubtedly recognised by other individuals as a signal of danger, it seems possible that this is not in fact their primary function. After all, while the animal making the display warns all others in the neighbourhood of the presence of potential danger, it also, in the process, draws attention to itself and perhaps increases its own risk. Nature is rarely so altruistic, for natural selection clearly acts at the level of the individual. Such self-advertisement could perhaps be justified on the basis of reciprocal altruism (Trivers, 1971) or, purely selfishly, in that the scattering of the rest of the group may distract or confuse the predator — with animals bounding about all around it. An alternative theory, however, is that the displays are in fact directed at the predators themselves, signalling 'I've seen you, you've lost the element of surprise' — with pronking perhaps adding the further information: 'Look how fit I am! I'm fast and healthy and not worth pursuing.' So far as any of these theories can be tested (in actually working out the costs and benefits attached), this last explanation seems the best one to account for the very similar behaviour of 'stotting' in Thomson's gazelle (Caro, 1986), and, in view of the fact that pronking in deer is not restricted exclusively to social species but is also recorded among roe and even musk deer, it seems probable that this is its major import among the deer, too.

While such an explanation for pronking and tail-flaring among deer had already been suggested by various authors (e.g. Bildstein, 1983), Caro's analysis in gazelle is the first time that such a hypothesis has been formally and objectively tested against all the other alternatives which have at various times been suggested. Caro in fact tested the validity of a whole range of different possible theories as to the function of stotting, concluding that hypotheses connecting the behaviour with attempts to startle or confuse a predator, or with invitation or deterrence of pursuit, were not supported by his data. Nor indeed was his evidence consistent with ideas noted above of the potential prey individual signalling its health. Caro concluded that all the characteristics of stotting in gazelle appear to be consistent only with a simple function of informing the predator that it has been detected. On the basis of such analysis, we may presume perhaps a similar function for the analagous behaviour shown by the various deer. We should note in conclusion, however, that alarm barks and these associated behaviours of pronking and tail-flaring are generally given by deer only when an animal is surprised or startled by a sudden threat or in response to a predator at relatively close range. More usually danger is recognised well in advance, by scent, sound or movement, and the deer move off swiftly and quietly before they are even detected.

5 Life histories and population dynamics

THE ANNUAL CYCLE

Many aspects of the behaviour and ecology of deer change in conjunction with changes in the animal's physiological state. Thus, the extra energy demands of pregnancy and lactation in females may influence habitat use and diet; stage of pregnancy or age of offspring may influence social behaviour and sociality in general. In the same way, habitat use, diet and degree of sociality in gregarious species, or intensity of territoriality in solitary species, may vary for males according to the stage in the sexual cycle. There are, for example, clear changes in social behaviour of males during the rut, during the period when antlers are shed and regrown, etc. Many of these physiological changes are periodic and vary predictably over a regular cycle. We have already described a physiological cycle of appetite and voluntary food intake in both sexes, which is clearly closely associated with changes in both diet and behaviour, and there are numerous other cyclic changes in physiology and behaviour which affect social organisation, patterns of resource use and many other aspects of the animals' lives. In considering in this chapter the life history of the different species of deer, therefore, it seems appropriate first to consider this annual cycle. (Since the factors affecting males and females differ substantially, we shall consider the sexes separately.)

The Annual Cycle in Females

In most deer species, females first become sexually mature from about 18 months of age (although in areas where food supplies are particularly rich and abundant they may mature more quickly: see page 102). From this time on they will start to show regular oestrus patterns. Among those species which live in relatively non-seasonal environments, oestrus may occur at any time of year, but, after the first calf has been born, the cycle settles to one of clear post-partum oestrus. In environments which show more marked variation in conditions at different times of year, there is commonly a much more distinct breeding season; all members of a population come into oestrus in synchrony, and the cycle is linked to changes in daylight pattern, rainfall or temperature. Oestrus cycles vary among species from around 14 to 24 days in length, and females will return to oestrus at regular intervals until they conceive. It is, however, normal for conception to occur during the first oestrus cycle. From this time on, most physiological changes in the female are associated with pregnancy, eventual birth of the offspring and lactation.

In solitary species, female territoriality increases in intensity as the time of parturition approaches and the females become even more solitary and secretive in habit until the young are a number of weeks old. Even among more social species there is a tendency for females to become more and more solitary as birth approaches. Individuals no longer form into the groups in which they may be encountered at other times, but remain entirely solitary. In the first few weeks after birth, and until they are strong enough to follow the mother easily, fawns are often left concealed in dense cover while the mother feeds; in those species characteristic of more open habitats, the mother may leave the fawn for considerable periods as she travels from the dense cover of fawning areas out to more open feeding habitats. At this time she may join up with other females into small groups while feeding, but returns on her own to where she has concealed her fawn(s). Only when the young are strong and fully able to stay with their mother will more permanent groups of mothers and their new offspring re-form. From this time on these 'small deer' groups are commonly encountered, and retain their identity right through the rest of the year. Female groups even stay together during the rut, forming a common harem, or visiting rutting stands as a unit; but the grouping tendency disappears once more as the time for birth of the next fawn approaches.

Figure 5.1 Seasonal fluctuations in the size of groups of 'small deer' (females and followers) among New Forest fallow deer. Source: Jackson (1974)

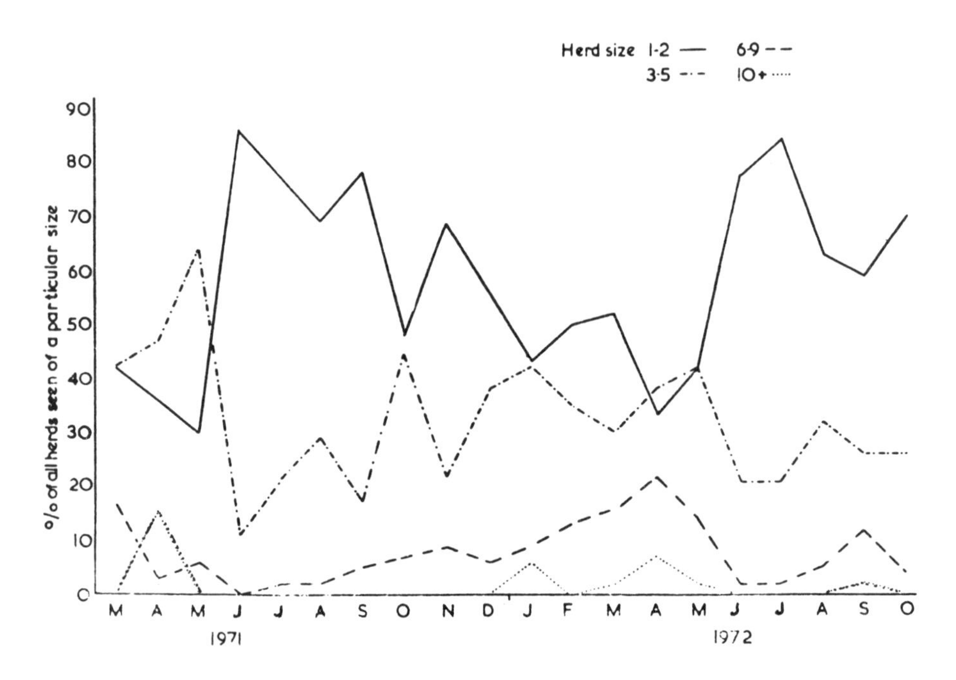

Such changes in group size of the more social species in association with this sexual cycle are clearly illustrated by John Jackson's observations of grouping in fallow deer in the New Forest in England (Jackson, 1974). Fallow are seasonal breeders, with the rut occurring during October and most births during June and July. Jackson's data (Figure 5.1) show clearly that there is a pronounced decrease in the number of large groups encountered during May and June, when females become more solitary as they prepare for the birth of their fawns; nearly 90 per cent of all observations are of solitary animals. By August, group sizes increase again as does and fawns join with other family groups, and rise once more during September and October as females collect at the rutting stands.

One other physiological change which occurs over the course of the year is the shedding of the coat. In animals of temperate areas, this is clearly synchronised with the change from winter to summer. The winter coat of coarse guard hairs and dense woolly underfibres grows into the summer coat during the autumn, from about September onwards, gradually replacing that thinner pelage. The new summer coat starts to grow in late spring, beneath the thick protective pelage of winter, which is shed during the spring after new growth is complete. In climates subject to less extreme seasonal variation in temperature, the moult pattern is less pronounced and parts of the coat may be replaced at almost any time.

Figure 5.2 Seasonal fluctuations in the group size of male New Forest fallow deer. Source: Jackson (1974)

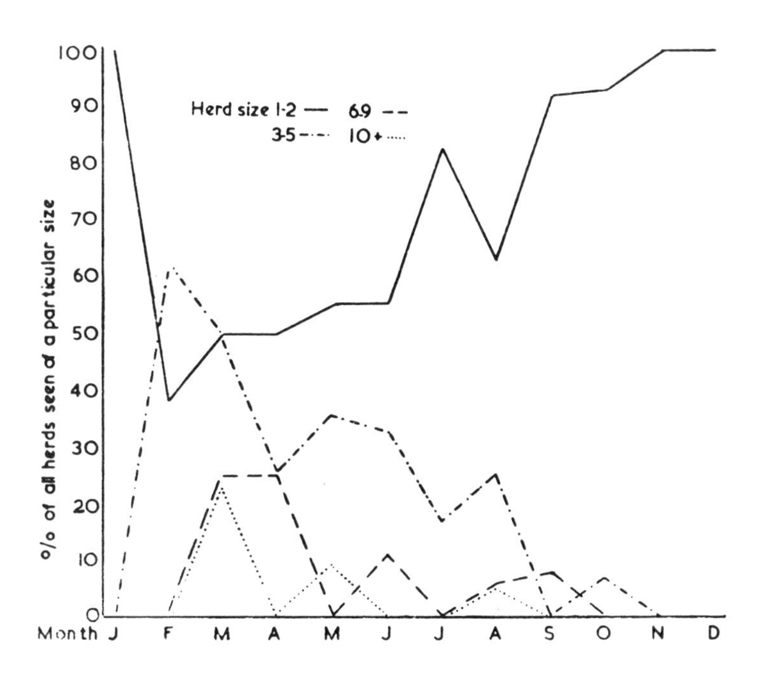

The Annual Cycle in Males

Since almost all physiological changes in the males of all species are associated with the hormonal changes of the sexual cycle, we may once again start our description of the annual cycle at the rut, that time when most active mating occurs. (In seasonal-breeding species this is closely synchronised, but even among aseasonal breeders each individual male, when actively involved in reproduction, is clearly at a defined point in his sexual cycle.) When in breeding condition, males of all species are characteristically at their most aggressive towards others. Territorial aggression is high among solitary species; contest for harems or rutting areas is intense among the more social species. While in most species viable sperm are present in the testes at almost all times of year, peak production coincides with the behavioural changes associated with active rutting, and both behavioural and spermatogenetic changes accompany a sharp rise in secretion of testosterone. Testosterone levels fall after the rut, and levels of aggression also diminish. Seasonally territorial species (such as roe and hog deer) abandon their territories and begin to form the small social groups characteristic of the non-breeding season; males of social species, too, begin to aggregate once more into bachelor herds.

After the intense activity of the rut (a period when, in many species, males are so involved in defence of territories, rutting stands or harems that feeding virtually ceases completely) the males are physically exhausted and need time to build up strength and reserves of body condition. All aspects of behaviour reflect this post-rut depletion of reserves: time spent feeding is higher than at any other period of the year; habitat use shifts towards those communities of highest food availability; aggression is at a low ebb; and group sizes in social species rise to a peak (Figure 5.2).

In those deer species whose males are antlered, however, one further change accompanies the cycle in testosterone secretion: antlers are cast during the 'slack' period of the reproductive cycle, and regrown in preparation for the next peak in competition. During this period males are vulnerable. We have noted that clear dominance hierarchies exist both in normally social and even in seasonally territorial species; among males, dominance rank relates not only to age and body size, but more specifically to antler size. A male which has shed its antlers rapidly drops in rank until these have regrown. During this time, the males become extremely secretive. Males of social species leave the bachelor herds for a period before shedding the antlers to become rather solitary, rejoining the groups only when growth of the new antlers is well advanced.

Moult patterns are also linked to this cycle of testosterone, but in temperate species shedding and regrowth of the coat is also entrained on the climatic seasons, with winter- and summer-coat changes occurring much at the same seasons as those of females.

SEASONAL AND ASEASONAL SPECIES

While annual cycles of this kind are evident in all deer species, and are accompanied by clear changes in behaviour of individual animals, only in species where the cycles of different individuals are strongly synchronised does the effect extend beyond the individual to influence the ecology or behaviour of the entire population. Males may become solitary for the period between casting and regrowth of their antlers; but in species whose cycles are not strongly synchronised we will merely observe individuals leaving and rejoining larger social groups of antlered males; the groups themselves will be apparent the year around, for there will always be some males in hard antler. By contrast, in species which show clear seasonality, all males shed their antlers at about the same time (although in practice there is some spread in that older individuals tend to cast before younger animals). In consequence, male herds more or less disband completely (e.g. Figure 5.2) and re-form when all animals have regrown their antlers. Similarly, in species where breeding is closely synchronised, sexual segregation outside the breeding season may be very obvious (page 61). If, however, some individuals — males and females — may be in breeding condition at any time of year such clear segregation of the sexes will not occur. Female groups at all times will contain one or two animals in a receptive state, and will always be accompanied by at least one male, an individual whose sexual activity is also at a peak at that particular time. Individual males join and leave the group at the peak of their sexual cycle, just as in more synchronised species, but in such aseasonal breeders there is likely always to be at least one male sexually active at any time of year.

In fact, as ever, we are describing extremes which rarely occur. Most species of deer are strongly seasonal breeders and their sexual cycles are thus closely synchronised: deer are, after all, predominantly a north-temperate group. Yet even here there is variation, and, while most female fallow deer for example will conceive in a brief three- to four-week period in late September and October, individual conceptions have been recorded as late as January (Sterba and Klusak, 1984). Equally, few species if any, are totally *a*seasonal. Even among those which are less strongly seasonal in habit, although conceptions, births, antler casting and regrowth can and do occur at any time of year, in practice there is still some degree of synchrony: there still remain times of the year when most conceptions occur or when most males have shed their antlers, and there is merely a wider variation in timing around those peaks. In his studies of four southeast Asian species in the Chitawan National Park of Nepal, Mishra (1982) demonstrated some degree of seasonality in chital, sambar, hog deer and muntjac. Although 'seasons' were more protracted than those typically observed in north-temperate species, and there were clearly individual animals so to speak 'out of phase' at any one time, antler cycle, male reproductive activity and calving showed some evidence of pattern. Figure 5.3 shows the seasonality of numbers of births for the four species, based on Mishra's observations, while Table 5.1 shows numbers of males recorded at different phases of the antler cycle in any month.

Arguably, these deer species, like those of the temperate zone, experience some seasonality in their environment, in this case between wet and dry seasons, to which the reproductive cycle might be entrained. Eisenberg and Lockhart (1972) suggested that, in Sri Lanka, breeding of chital is synchronised to seasons of drought and rainfall and consequently the availability of forage, much as in north-temperate species the breeding cycle is adjusted to the growing season. The fact that breeding 'seasons' for these animals and others, such as sambar, hog deer and

Table 5.1 Seasonality of antler cycle in four 'aseasonal' species (muntjac, hog deer, sambar, chital) expressed as proportion of males in each month observed at different stages of antler growth (data from Mishra, 1982)

Species	Antler stage	Month											
		Jan	Feb	Mar	Apr	May	Jun	Jul	Aug	Sep	Oct	Nov	Dec
Muntjac	% hard antler	100	100	78	33	31	27	18	42	80	93	100	100
	% cast	–	–	9	19	9	–	–	–	–	–	–	–
	% in velvet	–	–	13	48	60	13	82	58	20	7	–	–
Hog deer	% hard antler	50	57	18	18	65	92	93	85	100	100	67	60
	% cast	–	6	14	–	–	–	–	–	–	–	17	–
	% in velvet	50	37	68	82	35	8	7	15	–	–	17	40
Sambar	% hard antler	92	87	88	81	55	42	12	14	27	33	67	91
	% cast	–	2	3	7	16	3	13	9	–	–	–	4
	% in velvet	8	11	9	12	29	16	75	77	73	67	33	5
Chital	% hard antler	13	20	46	89	98	100	100	90	74	45	28	10
	% cast	4	2	1	–	–	–	–	7	12	18	20	17
	% in velvet	83	78	53	11	2	–	–	3	14	37	52	73

Figure 5.3 Season of births among populations of muntjac, sambar, hog deer and chital in the Chitawan National Park in Nepal (data from Mishra, 1982)

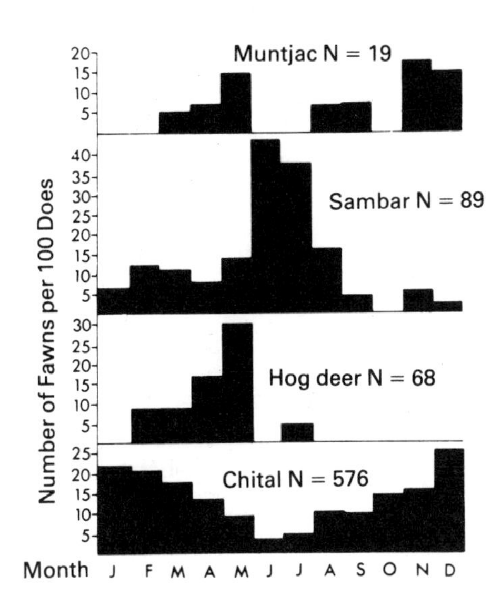

muntjac, are not consistent throughout their range adds some support to this, suggesting that this variation itself may relate to regional differences in climate and vegetation (e.g. Prater, 1945; Krishnan, 1959; Berwick, 1974). Few, if any, deer species experience absolutely no seasonal change in their environment. Selection will act to ensure that most births occur during that season of abundant forage. Perhaps differences in the degree of synchrony of the annual cycle in different species thus merely reflect the degree of contrast apparent between the different seasons experienced, or the absolute length of the restricted season of active forage growth. With greater seasonal contrast apparent in more northerly latitudes, and shorter seasons of abundant vegetative growth, individual variation in breeding cycles may be expected to be less, resulting in the tighter synchrony in the annual cycle of the species characteristic of such areas.

Further support for such a suggestion that breeding cycles must, throughout, reflect the periodicity of vegetative production (being thus more closely synchronised in more northerly species) may derive from the observation that the cycles of different species in any one area tend themselves to be well synchronised. Thus, in north-temperate regions, most species rut in the autumn, with fawns born in the spring. In Europe, the peak rutting seasons of red deer, sika deer and fallow deer all occur between the end of September and the end of November; young of all three species are born between the end of May and mid-June, with a few stragglers continuing to produce live offspring up until the end of July; males of all species cast their antlers in April/May and growth of the new antlers is complete by late August, ready for the next rutting season. The cycles of all species are essentially extremely similar. The only exception to the rule is the roe deer, in which males cast their antlers during October and November, with regrowth complete for a summer rut in late

June and July. Roe deer, however, have a rather curious breeding system in which implantation of the embryo does not follow directly after fertilisation but is delayed until December or January, and only then does the foetus begin to develop. As a result roe drop their young in May, much like the other European species — at the appropriate season of abundant plant growth.

We should not, however, get too carried away with these nice theories explaining the greater apparent seasonality of northern species in terms of the need to concentrate births of young into a relatively shorter season of active plant growth. What is particularly interesting in such a context is to look at the degree of seasonality demonstrated by a relatively cosmopolitan species such as the white-tailed deer over different parts of its range; or to examine what happens to the cycle of relatively less well-synchronised animals such as muntjac or other south Asian species when they are introduced into countries with more strongly seasonal environments. In such a context, Figure 5.4 summarises the seasons of birth for eight species of deer, of various origins, kept at London Zoo (from Lincoln, 1985). Although it is perhaps slightly unfair to extrapolate too far from behaviour of what are of course captive animals, it should be apparent that species such as the chital, hog deer and sambar reveal no great seasonality in breeding, even after translocation to a temperate climate. Data on birth dates of wild (feral) Chinese muntjac in the UK, collected by Chapman *et al.* (1984), which show that fawns may be born at any time of year, also suggest that the lack of seasonality in such species is inherent and not simply related to lack of any marked environmental cycle in their native lands.

Figure 5.4 Season of births for eight species of deer kept in the London Zoological Gardens (latitude 51°30′N) between 1834 and 1937. Species are ordered according to the latitude of origin. Source: Lincoln (1985)

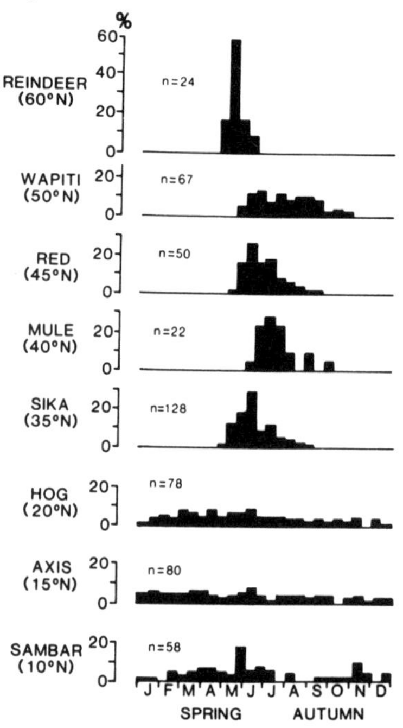

We have in these last few pages reviewed something of the cycle of change in the lives of deer over the course of a year. It seems appropriate that we should now move to the longer time-scale and describe the fuller *life* cycle.

THE LIFE CYCLE

For most species of deer, as we have already noted, the breeding season is fairly concentrated, with females synchronised to come into oestrus over a relatively short period of time. We have shown that, even for less markedly seasonal species, most reproductive activity is still concentrated into a comparatively short period; it is merely that this period is somewhat more protracted, a matter of months rather than weeks.

Mating systems among solitary and territorial species are straightforward. As we have noted, males and females of such species establish their territories independently of one another; in the appropriate season males mate with those females whose territories overlap with their own. Such a bald description is somewhat misleading, however, for it gives the impression that the interaction is essentially 'casual' and restricted merely to copulation by males of receptive females. Observations of both roe deer and muntjac suggest that this is not quite the case. Males may form quite close associations with the females within their range, and are commonly encountered consorting with them, whether they are in oestrus or not. Mutual grooming is commonplace (to the extent that radio-telemetric collars or ear-tags fitted to muntjac by research scientists for recognition purposes are well known to be frustratingly short-lived!). One might suggest that quite a strong bond may thus be formed between partners in these species; after all, they share the same home range all year around and have in fact far more opportunity for contact than the so-called 'more social' species, whose sexes may be segregated for much of the year.

Mating systems among these latter species are themselves more complex. We have noted that, with greater 'sociality', there is a greater tendency towards polygyny, and indeed have suggested (page 29) that this would necessarily follow. Polygyny can, however, be accommodated within a number of actual mating strategies: the collection of harems; foraging among 'unowned' female herds for oestrus females; the establishment of a rutting stand or display ground to which to attract females; or, admittedly more rarely, communal display in a form of lek. The system adopted differs widely between species, and between circumstances. In species in which breeding seasons are protracted, then establishment of rutting stands or leks would perhaps be inappropriate. Maintenance of harems becomes an expensive strategy, for these must be held and defended against other males for very considerable periods, not merely the few weeks of synchronised oestrus of seasonal breeders. Harems are formed and held by some species, but more commonly adult males forage among the herds for oestrus females. Males may join the herds when sexually active, leaving again when their peak is past (as, for example, sambar or rusa deer), or they may run permanently with the females in mixed groups, competing with other males for oestrus females during that period when they themselves are sexually active. In such systems clear hierarchies are established among the males, with dominant males accomplishing most of the matings; but dominance ranks change as individuals come into full breeding condition or pass into 'eclipse'. Furthermore, the strain of competition for females is sufficiently severe that no male remains dominant for long. Dominant males must constantly repel the challenge of others; continuously fighting and with little opportunity to feed, the master stags rapidly lose body condition and initial dominance is short-lived.

Where breeding seasons are more closely synchronised, the options are more varied: males may hold display territories or compete for ownership of groups of females. The strategy adopted seems, however, to vary with circumstance, and any one species may show a whole range of responses. Red deer are normally harem breeders while fallow, for example, are a species which characteristically holds rutting stands. As mentioned earlier, however, the whole social organisation of fallow deer is flexible and seems to differ markedly with changing environmental character; correlated with that change is an appropriate change in reproductive strategy. In woodland areas, or areas of fairly continuous closed cover, fallow deer form clearly distinct male and female groups segregated even to the degree of occupying distinct geographical ranges. Males move into the female areas and set up rutting stands to which they attract females; the bucks make only slight attempt to retain females on their stand after copulation, and the amount of female herding is small. In more open areas where woodlands are perhaps only small copses scattered among agricultural land, males also enter the females' areas in the autumn to set up rutting stands. Once a group of females has been attracted to the stand, however, and perhaps because in such environments deer are anyway likely to be at much lower densities, the male switches the focus of his attention far more towards his does than his actual display ground. He herds them regularly and solicitously; and even after the main rut is complete he remains with them, leading his harem to and from favoured feeding grounds, returning to the male range only when the harem itself disperses in late spring as females prepare for the birth of their fawns. In really open country, where herds containing adults of both sexes may be observed all year round, males forage for oestrus does within the herd, competing among themselves for dominance and thus the *droit de seigneur*. Even among the conventional rutting-stand holders there is room for variation, and when available sites for display grounds are sparse, when male density is high or females highly aggregated in a few areas, rutting fallow bucks may come together to form a communal lek (pages 72–3). Similar variation is reported in, for example, sika deer, which in some situations seem to hold harems, but in others mark and defend distinct breeding territories. Once again the strategy adopted seems to depend on environmental circumstances, and on both the density and the spatial distribution of females.

Whatever the system adopted, males of polygynous species must compete: for display ground or for females. They must advertise to attract females and, if a 'territory-based' system is adopted, must mark and defend their patch of land.

The Rut

Contest between males for dominance within a mixed herd, for sole possession of a harem, or for a display ground, is intense. While much is achieved through display, fighting is commonplace. In display males 'puff up' to their full size, call loudly at each other in challenge, and paw and scrape at the ground. Antlers are held erect and 'paraded', and, as noted earlier, antler size and body size itself are closely related to dominance. A male may engage in a 'mock fight' with nearby vegetation, thrashing with his antlers at trees or thrashing them into the ground vegetation; quite commonly as the result of this, antlers end up 'decorated' with great tangles of vegetation, and it is quite possible that some species deliberately gather vegetation upon the antlers in this way to increase their apparent size. If disputes cannot be resolved by display of this kind, antlers will be interlocked and rival males engage in pushing and wrestling contests as described on page 84.

Victorious males must attract females: in those species for which ownership of, or access to, oestrus females is the prize for success this is less apparent, but where mating rights accord with ownership of land females must be brought to the rutting stand. Advertisement is by sight, sound and scent. The male may mark the area of his stand by scoring the bark of nearby trees, by thrashing vegetation or making scrapes in the ground. Scrapes are scented by the secretions of interdigital glands — though this often seems rather superfluous, for a rutting male can in any case be scented over a very considerable distance! At the peak of his sexual cycle a male's suborbital glands and preputial glands secrete actively; the male will also anoint himself by spraying his own body (flanks and legs) with urine. Male deer of most species occasionally seek out mud-wallows and roll in them, for the cooling effect in summer, or perhaps to counter the irritation of external parasites; but use of wallows increases dramatically during the rutting period, and the odour cues may be enhanced by the buck urinating into the mudhole before wallowing. Finally, the male may attract females to his rutting stand by calling: the same call as is used in aggressive display against rival males. It has been observed that the quality of the call provides an accurate reflection of the strength and dominance of the caller (Clutton-Brock and Albon, 1979), clearly a useful cue in aggressive display; that same correlation between quality of call and quality of caller can, however, also be used by females to select the best mate. Males of all polygynous species call during the rut. In harem-holding species the call is essentially merely male competitive display; in stand-holding species the call serves this double function of male challenge and female attraction. Calls vary tremendously, from the high-pitched whistle of an adult sika stag to the roar of a red stag and the deep bellow of a wapiti bull or a bull moose; fallow deer belch or groan. Whatever the actual tone, however, the call always has a curiously carrying quality.

Actual courtship is fairly perfunctory: males sniff the vulva of potential mates, either in their harem or attracted to their display ground, to detect those in oestrus. As with males of all ungulate species (and many other mammals), 'testing' of the scent is accompanied by the lip-curling 'flehmen' by which the openings of the highly sensitive Jacobsen's organs (olfactory pits which open into the soft mucosa of the inside of the upper lip) may be exposed to the scent. Receptive females are then gently encouraged. Males may pursue them in short chases, or walk behind them continually nuzzling at flanks and vulva, until they stand to be mounted. Commonly, a male will mount a female a number of times before intromission and ejaculation.

In those mating systems based on ownership or dominance of a rutting stand, the actual stand is often in a traditional site: it persists from year to year in the same place — far longer than the length of time it may be held by any one male. The stand itself is traditional; it is not merely that one male will develop a site and then use the same site in successive years, but rather that males compete to occupy the traditional stand. A given male, however, may indeed occupy the same stand for a number of seasons before he is overthrown. Equally, females will tend to return year after year to particular stands, bringing their offspring with them to the same stand. And therein lies a curiosity: although the period over which any one male may win and hold a given stand from year to year is limited, and it would be rare for the same buck to stand on the same site for more than perhaps two or three years, during that time, since females, too, remain faithful to a stand, he must indubitably mate with the same does each year; further, since those females are accompanied by their female offspring of previous years, the latter stand a good chance in the first few years of their reproductive life of being mated by their own father. The implications of the inevitable inbreeding that results from such a system are examined for fallow deer by Smith (1979).

Pregnancy and early development

Females usually conceive on the first oestrus cycle of any breeding season. If not, in the majority of species, they will return to oestrus at regular intervals until they do conceive. The length of the oestrus cycle in such polyoestrus species varies from about 18 days (e.g. red deer, sika) to 21 days (e.g. fallow deer). Gestation periods vary between about 180 and 240 days: usually, the larger the animal the longer the period of pregnancy. Thus, in Chinese water deer gestation may be as short as 180 days, while in red deer pregnancy may continue for as long as 240 days. The longest *apparent* gestation period is that of the European roe deer, in which the young are born some ten months after conception. Such a protracted pregnancy is, however, somewhat misleading, for in roe deer implantation of the embryo into the uterine wall does not follow immediately after fertilisation but is delayed about five months. After fertilisation during the summer rut (July-August), the blastocyst increases only very slightly in size and remains free within the uterus until late December or early January. Only then does the embryo become attached to the wall of the uterus and begin to grow. From this time on growth is extremely rapid, with kids born in May or June, some 21 weeks later. Roe are the only species of artiodactyl known to practise delayed implantation in this way, and the function is obscure.

Among the more advanced Eurasian deer, the Cervinae, multiple births are uncommon and most species give birth to single offspring, although there are rare exceptions: e.g. for fallow deer, Harrison and Hyett (1954), Baker (1973); for red deer, Guinness and Fletcher (1971), MacNally (1982), Ratcliffe (1984). Smaller and more 'primitive' species tend to be more prolific: in muntjac and musk deer, twin births are relatively common, and Chinese water deer regularly produce twins and triplets (Middleton, 1937), with litter sizes of up to five or six recorded in China (Hamilton, 1871). This apparently primitive character, however, seems to be lost in the more advanced Eurasian deer. By contrast, the trait is retained to an extent among some members of the other 'advanced' group, the Odocoilinae. Roe deer and moose for example regularly conceive twins or, more rarely, triplets, and multiple conceptions are also common among white-tailed deer and mule deer. (Harder, 1984, for example, for one population of white-tailed deer in Ohio, estimated conception rates in mature does ranging from 1.56 to 1.87 in different years, with a five-year average of 1.67.) Not all foetuses conceived, however, will necessarily be developed full-term; in all species the numbers actually born seem to be related to population density, with twins more common in low-density populations, or those with richer food supplies, and only single offspring born in areas of higher density.

Unlike young antelope, which are able to get up and run with their mothers within only a few minutes of birth, baby deer are relatively helpless for the first few weeks of life. The infant coat is usually spotted for camouflage, and the young deer remains hidden in a 'form' in dense vegetation for most of the day while its mother forages, returning occasionally to suckle it (Plate 23). Where a mother produces more than one fawn, these are commonly established in different forms some distance from each other (presumably to reduce the concentration of scent which might otherwise increase the risk of predation) and the mother moves regularly between the two (e.g. for white-tailed deer: Downing and McGinnes, 1969; White *et al.*, 1972; Hirth, 1973). Within a period of a few days to a few weeks, however, the infant is strong enough to join its mother and accompanies her from that time on.

Smaller species tend, not unnaturally, to mature more rapidly than larger ones; accordingly, the young of these smaller species generally start to take solid food early and may be completely weaned within two to three months. In larger species, the young continue to suckle for much longer. Of course occasional suckling attempts may be made by quite mature calves of almost any species, but regular suckling of young by wild fallow deer has been frequently observed as much as six months after parturition and 90 per cent of New Forest fallow deer females culled between 1971 and 1973 were still lactating after seven months (Jackson, 1977). Lactation has also been monitored in red deer and is equally protracted, with milk production peaking around the seventh or eighth week after birth and declining after the fifteenth week, but with susbstantial production still maintained even after 200 days (seven months) (Arman *et al.*, 1974). Lactation clearly represents a considerable drain on the mother: for Scottish red deer, Arman *et al.* (1974) calculate yields of 1.4–2.0 kg per day during the peak of lactation; Loudon and Milne (1985) estimated a mean daily production of between 1.2 and 1.7 kg maintained over the first 100 days of the lactation; and these figures are low by comparison with Bubenik's (1965) estimate of 3–4.5 litres per day for Continental red deer. Even though birth and lactation are usually timed to coincide with periods of peak food availability, females are frequently not able to take in sufficient food of sufficient quality to meet the increased demand, and must draw on bodily reserves of energy; in consequence, condition of mothers commonly falls, at least during early lactation, and recovers only slowly in the later weeks of milking.

In sexually dimorphic species male calves are usually larger at birth, and certainly grow more rapidly than females in the early weeks of life.

Demands on the mother thus tend to be even higher for male offspring than for female calves: in their studies of red deer in Scotland, Clutton-Brock *et al.* (1979) showed that male calves suckled more frequently than females and also tended to suckle for longer. Similar results have been recorded for caribou: male calves are born heavier than females, grow faster and have a higher milk intake (McEwan, 1968; McEwan and Whitehead, 1970).

In theory, the difference in investment in sons and daughters in such species should not merely be a passive response to the greater demands of a larger, male, calf. Sexual dimorphism is at its most marked in those species with polygynous breeding systems, because in these species male competition for females will be at its most intense (page 27). Yet in such species, while all females will have an equal chance of producing offspring, males will succeed in fathering young only if they are dominant and can outcompete other males. There are distinct 'haves' and 'have-nots' — and dominance is related to body size and condition. If we consider, in evolutionary terms, any individual's 'fitness' in terms of the number of offspring left to succeeding generations, whether its own offspring or their descendants, then, for any male offspring born, a female will obtain deferred reproductive success in future generations only if that son ends up as a dominant animal who will secure mates. A male offspring who never achieves dominance is a 'wasted' reproductive effort in terms of long-term perpetuation of the genes. It can be shown that the ultimate size, and thus dominance rank, achieved by an animal is closely correlated with its rate of growth in the first year of life (Clutton-Brock *et al.*, 1982).

Arguably, therefore, females should actively invest heavily in any male offspring born, so that those male calves may have the greatest chance of ultimately ending up dominant, and thus reproductive, adults. Such investment would not only be reflected in increased milk yield provided for male rather than female calves, but might be expected to be reflected also in a more general increase in the degree of maternal care given. Not only might male calves be suckled more frequently and for longer periods in any bout, but their suckling might be continued over a longer duration overall (i.e. weaning delayed); mothers might be expected to spend more time in contact with male offspring and more rarely to reject contacts initiated by the calf itself. There appears to be evidence in support of such a hypothesis among the red deer studied by Clutton-Brock *et al.* (1981; 1982); where similar studies have been undertaken of other species, e.g. fallow deer (Boddington, 1985; Gordon, 1986; Gauthier and Barrette, 1986), and in white-tailed deer (Gauthier and Barrette, 1986), however, no differences of any kind have been recorded between the behaviour of mothers with male offspring and that of those with female fawns.

Let us pursue the argument a stage further. While female offspring are comparatively less expensive than male offspring, and will, when mature, themselves produce one offspring every year, a male calf in polygynous species, *if successful*, may father many offspring in each year in which he is reproductively active (and if this is, let us say, seven or eight young per year over a reproductive span of even three to four years, it represents a greater lifetime reproductive achievement than a daughter's production of one calf a year over perhaps eight to ten years).

To ensure a male's success, however, his mother must invest heavily in him, perhaps reducing her own chances of breeding in the following year, or even reducing her own life expectancy. What should she do in such a situation? Clutton-Brock *et al.* suggest that ideally, while a female is herself young, she should preferentially play safe and produce female offspring. Rearing of female calves will not significantly reduce her own future lifetime reproductive success, either by shortening her lifespan or by threatening her fitness to conceive in future years, and they will themselves produce her at least one grandchild every year of their active adult lives.

As a female grows older, however, the balance of risk changes. Her own probability of survival from one year to the next decreases, and there is no guarantee in any one year that she will herself live to breed again in the following year. She may, but she may not; her own personal lifetime reproductive output is nearly complete. To maximise her long-term contribution of genes to the next generation now, her best 'bet' is to use her last remaining reproductive potential in producing strong male calves, and to invest all the extra cost required to ensure that they will ultimately grow to become dominant reproductive animals likely to succeed in competition for mates. She takes the risk that she may in consequence never reproduce again herself, but then her probability of surviving to the next year is in any case low; she is in addition more experienced in rearing calves and because of this can do so at lesser cost to herself — reducing the influence on her own future reproductive output — and with greater success. (Calves of experienced mothers always survive better and grow more strongly than those of primiparous females; male calves are thus more likely to become dominant as adults.) By investing now in male offspring, therefore, she would with her own last, or last few, reproductive events maximise their eventual impact on later generations.

A nice idea in theory? Surprisingly, Clutton-Brock's studies of red deer show that there is indeed a tendency for mothers to produce female calves at the beginning of their reproductive life, but to give birth to a greater proportion of male calves as they themselves grow older (Clutton-Brock *et al.*, 1982).

One consequence of investing heavily in offspring, of whatever sex, is that it draws too heavily on the mother's own energy reserves, and may perhaps mean that she does not build up condition sufficiently rapidly to conceive in the next rut. Normal inter-calving intervals among seasonal breeders provide for conception in each breeding season, and thus adult females produce a calf once a year. Among less strongly seasonal breeders, females often conceive on a post-partum oestrus: inter-calving intervals of, for example, muntjac (admittedly in captivity, and see page 94) are recorded as 210 days (with gestation 200 days). While such regular production of young may be maintained in areas of good food availability, however, or where population density is relatively low, when food is in shorter supply females may not recover body condition sufficiently quickly to conceive at the normal time. In 'aseasonal' breeders this may just delay conception until a later date in the same year, but in seasonally breeding species a female will in consequence miss a year altogether. Thus, at any one time, one may find within a population females too young to breed, others adult and pregnant or

Figure 5.5 Probability of conception is related to female body weight. The graph shows the relationship between the probability of calving and female body weight at the time of the rut for a group of farmed red deer hinds (based on data from Hamilton and Blaxter, 1980)

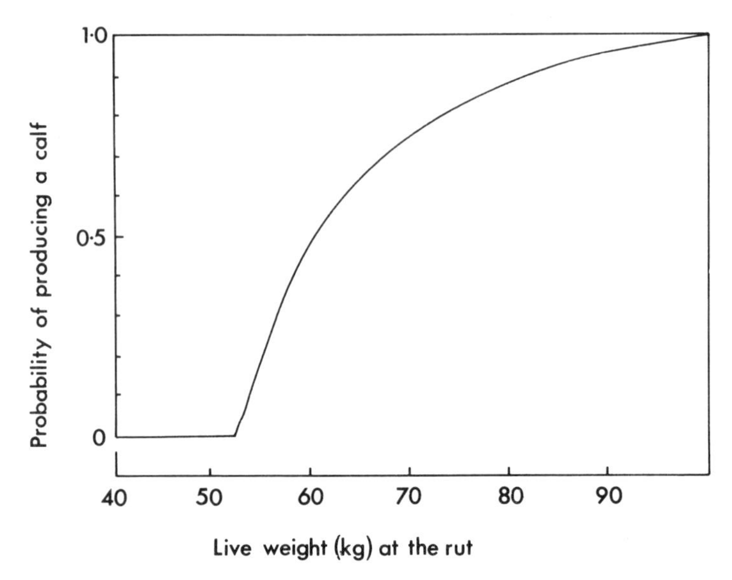

lactating, and others again, adult and capable of reproduction, but currently non-reproductive. These last are commonly referred to as being 'yeld'.

The explanation of such a phenomenon is quite simple. Ability to conceive is related not simply to age, but to body weight, and unless a female reaches a particular threshold weight she will not ovulate. This appears to be a widespread phenomenon and is recorded in, for example, roe deer, fallow deer, white-tailed deer and red deer (Joubert, 1963; Mitchell and Brown, 1974; Staines, 1978; Ratcliffe and Rowe, 1985; Teer, 1984). Figure 5.5 shows the number of farmed red deer hinds successfully conceiving at different body weights, and makes it clear that such hinds must reach a weight in excess of 60 kg if they are to have a realistic chance of conception. In populations of low density, or where food is very abundant, weight gain is rapid. An adult female can recover the weight loss during early lactation quickly enough to reach the critical body weight in time for the next rut. In areas of low overall food availability or intense competition for food, however, weight loss during lactation may not be so quickly recovered.

Maturity and Dispersal

This same relationship applies in determining age of sexual maturity. After weaning, deer calves continue to grow steadily towards adult weight (although, Figure 5.6, there will be a check in growth during winter in temperate species: calves, like adults, have a period of winter inappetence). Growth rates vary with environmental conditions and, once again, body weight is more important in determining sexual maturity than actual age. In different areas, females of the same species may be found to be pregnant as yearlings or even, in exceptional cases, as fawns, or alternatively may not reproduce for the first time until they are two years old or more. In each case it appears that this is the length of time taken in the different areas to reach the critical body weight for ovulation; in areas where food is of good quality and availability, females will reach that threshold weight at a younger age.

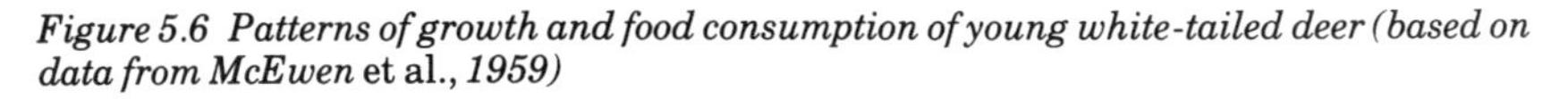

Figure 5.6 Patterns of growth and food consumption of young white-tailed deer (based on data from McEwen et al., *1959)*

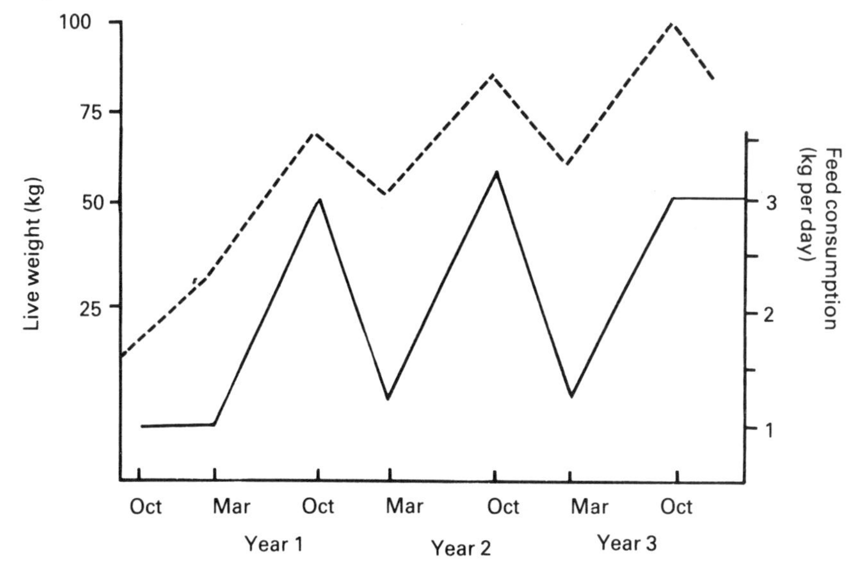

The age at which males first reach sexual maturity is also related to weight, though to a lesser degree. Testes become active in hormone production towards the end of the first year of life, and the first antlers are grown during the second year. Active spermatozoa are present in the testis also from this time, but a male will of course not breed for the first time until he is sufficiently old and high-ranking to win a territory or, among more social species, to compete successfully with other males for access to oestrus females, and this may not be for a number of years.

Even after they are no longer suckling and are fully weaned, young deer remain dependent on their mother for a considerable period. Fawns of territorial species usually remain in the territory of their mother until she is about to give birth to the next season's young. In seasonally territorial species such as roe deer, fawns remain with the mother through the break-up of territoriality in autumn, and the groups encountered in the winter (or equivalent non-territorial season in other species) are usually amalgamations of such family parties. Only when territories are re-established in the following year are the yearlings driven out. In most cases, too, females are far more tolerant towards their female offspring and daughters are quite regularly permitted to remain within, or on the edge of, the mother's range; among such territorial species, where overlap of ranges is observed, it is usually found to be due to overlap between ranges of mothers and their daughters (e.g. Hirth, 1973; Loudon, 1979). Males are, however, vigorously rejected, and must disperse from the natal area in search of a territory of their own. Of course, at this time, adult males are also redefining their territorial boundaries, and are equally intolerant of the dispersing juveniles. The juveniles must establish a territory or die — but there are few vacancies. Yearling males may be forced to travel a long way from their natal area (roe deer have been recorded over 48 km from the place they were marked as fawns); in such journeys again there is high risk of mortality, through accident, or merely owing to the fact that it is harder to forage in unfamiliar country; here, again, they will be subject to harassment by the local territory-holders. As a result of all this, mortality is incredibly high: yearling mortality among males of such territorial species is commonly of the order of 90–95 per cent.

Yearling males of social species are also eventually forced to leave their mother's range. Fawns of both sexes will join the mother's social group when she returns to it after her 'confinement'. They will accompany her during the rutting season that follows their own birth, when harems, or groups at rutting stands, are regularly accompanied by the current year's fawns. Commonly they will also stay with her when she leaves the herd again the following year for the birth of her new fawn, but from this time on she increasingly rejects any approaches to suckle, and terminates all actually maternal behaviour. Interactions are restricted purely to the normal social behaviour shown to any other individual of the group. As female fawns assume maturity, they may, like those of territorial species, remain within the mother's range (indeed more usually do so) and frequently stay on as independent adult members of the 'social clan' (page 60). Among species or populations in which social groups are fairly constant in composition and not as fluid as would more normally appear the rule, this may result in the development of clear matrilineal groups (pages 60, 70). Males again, however, are not tolerated indefinitely and, either on their first birthday, or more commonly at the time of the second rut following their birth (by which time they will be about 16 months, bearing their first antlers, and themselves becoming sexually active, certainly enough to pester their mothers!), they will be driven out, or drift away from the female groups to form their own bachelor herds or join with more mature males of already established male groups.

Mortality

We have noted an enormously high mortality associated with dispersal of juvenile males in territorial species of deer; such levels may be extreme, but heavy mortalities occur during the first year of life in all species, and among female as well as male calves. In the red deer populations studied by Clutton-Brock *et al.* on Rhum, 20 per cent of all calves died before the end of the September following their birth and a further 11 per cent died in the following winter. Winter deaths usually occurred towards the end of the season, and the proportion dying increased with both severity of the weather and population density. In addition, late-born calves were found to be more likely to die than calves born early in the season (Guinness *et al.*, 1978; Clutton-Brock *et al.*, 1982). For American wapiti, Houston (1982) estimated mortality during the first six to nine months of life as between 50 per cent and 70 per cent, and figures quoted for other species are of the same order of magnitude (e.g. white-tailed deer 42 per cent: Eberhardt, 1969; fallow deer 50–60 per cent: Chapman and Chapman, 1975).

Table 5.2 Survival expectations at different stages of life for white-tailed deer (data modified from Hayne, 1984). Each value in the table shows the number of animals of a given age which remain alive as a proportion of the number of those same animals which were alive the previous year, and in effect represent the survival prospects for animals of any given age from each year to the next

Age in years	Proportion of animals of age (x – 1) years surviving to x years	
	Females	Males
1	0.58	0.58
2	0.70	0.28
3	0.70	0.28
4	0.70	0.28
5	0.70	0.28
6	0.70	0.28
7	0.70	
8	0.70	
9	0.70	
10	0.70	
↓	each year 0.70	
20		

Beyond this critical period, patterns of mortality level out, and absolute levels of mortality are actually comparatively low until, towards the end of the animals' lifespan, they increase again with old age. Life tables, showing the probability of survival from one year to the next among female white-tailed deer (Table 5.2), show that, after the initial high mortality in the first year of life, mortality rate declines and remains relatively constant over the middle years. Figure 5.7 reveals for the red deer population on Rhum an essentially similar pattern of high initial mortality, with a phase of relatively high survival over the middle years and then an increase in mortality once more among older animals.

One further important point is apparent in both Table 5.2 and Figure 5.7: it is clear that, although levels of mortality in the first critical year are much the same for males and females, survivorship of those individuals which have lived beyond their first year is not equivalent for the two sexes. Mortality of males is far higher than that of females and overall life expectancy is far shorter. Among territorial species, males tend to hold larger territories than do females (page 77) and are less tolerant of any range overlap. Accordingly, it may be harder for males to hold and retain a territory, and thus we might perhaps expect mortality to be somewhat higher than that of females. The pattern of increased adult mortality of males over females, however, is also preserved among more social species: indeed, in such species males suffer considerably higher levels of mortality than females, and the imbalance tends to become even more marked at high population densities (e.g Robinette *et*

Figure 5.7 The population structure of red deer on Rhum. The figures present the percentage contribution made to the total population by each sex- and age-class. (a) Year-class percentages on 1 June 1957 (from Lowe, 1969); (b) year-class percentages in one population in 1971. (Source: Clutton-Brock, et al. *1982) It is clear that, after heavy mortality early in life, levels of mortality drop and are at a constant low rate among mature deer. Note that there is also a difference in mortality schedule between the sexes*

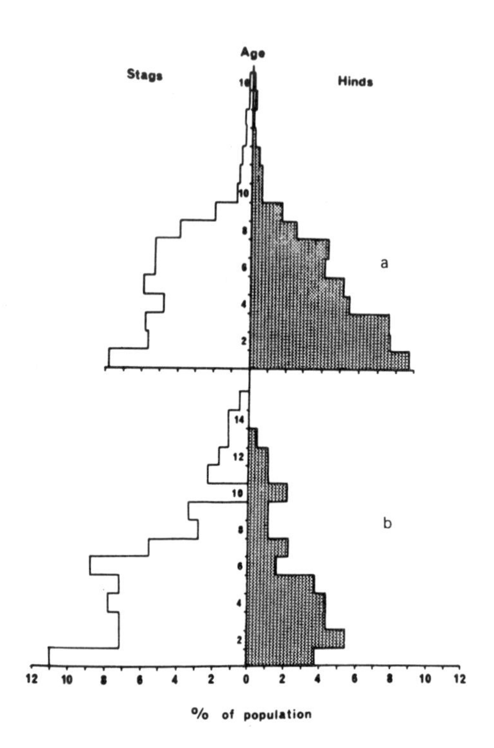

al., 1957; Klein and Olson, 1960; Clutton-Brock *et al.*, 1982). Possible reasons for this imbalance are explored in more detail below.

Among mammals in general, lifespan increases with body size and deer reflect this same general trend. Average lifespan is a difficult statistic to calculate or interpret, since such a high proportion of animals die within their first year of life anyway, and after that time mortality rates of those which do survive are usually lower by comparison. The maximum ages recorded for a number of species may, however, be noted here by way of example: muntjac in the wild have been known to live to 10–14 years; roe deer marked in the wild as fawns have been observed up to 16 or 17 years later; sika deer known to the author have been recorded at more than 20 years old; and red deer have been reported at up to 26 years. These are of course all extremes and, as Figure 5.7 makes clear, even if they survive their first year, few red deer indeed would actually survive beyond eight years. Some comparative data are available for wild sambar and chital in the Ruhuna National Park in Sri Lanka. Here Ashby and Santiapillai (1986) estimate for chital deer a maximum life expectancy of 14 years, while for sambar, although they may live to a maximum of 24 years, mean life expectancy was lower than this, at approximately ten years.

Causes of Mortality

Death during the first few months of life is usually due to rejection by the mother, accident, or predation. Overwinter mortality or death during dispersion is almost invariably due to malnutrition. Causes of death in mature animals are more varied. Disease and accident indubitably account for a number of animals. While wild deer are in fact relatively resistant to disease, a high proportion carry a substantial burden of both internal and external parasites (e.g. Chapter 6, Table 6.1). While these in themselves are unlikely to cause direct mortality (except in the case of the lungworms, *Dictylocaulus* spp., which may give rise to parasitic pneumonia, an important cause of overwinter mortality of young roe deer), they may lower condition, making the animal more susceptible to temperature, malnutrition or predation.

Among adult males, fighting for territories, rutting stands or harems is an additional source of mortality. Clutton-Brock and his co-workers estimate that on Rhum some 25 per cent of all red deer stags over the age of five years show some sign of injury during the rut each year, and up to 6 per cent were permanently injured; since most stags rut for perhaps three to five years during their lifetime, this suggests that as many as 20 per cent may sustain permanent injury during their life. The frequency of rutting injuries is known to be high in other deer populations. In two samples of Russian red deer, mortality from rutting injuries accounted for 13 per cent and 29 per cent of all adult male deaths (Heptner *et al.*, 1961). In a sample of mature mule deer observed over a single breeding season, 19 per cent showed some sign of injury (Geist, 1974), while in reindeer and moose rutting mortality is also a common cause of death in adult males (Bergerud, 1973; Pielowski, 1969). Since even slightly injured individuals probably run a considerably higher risk of winter mortality, the real costs of fighting in terms of mortality may be even higher than these figures suggest (Clutton-Brock *et al.*, 1982).

Such a high incidence of injury during competition for mates seems, intuitively, to offer a ready explanation for the fact that mortality levels among adult males even of non-territorial species are greater than those for females of the equivalent age. In practice, however, it seems improbable that the difference in mortality between the sexes can be attributed entirely to this. Much male mortality in such species occurs before the animals reach an age at which they would even start to compete for females, and it seems likely that a major contribution to the increased mortality of males is related more to the dimorphism of size associated with polygyny rather than to the polygyny itself. Because they are larger, males have in general greater nutritional requirements than females and may be less well able to fulfil them in conditions of poor food availability. We have seen (Chapter 3) that this in itself may lead males to occupy rather different habitats from those occupied by females, and to feed in general on a greater bulk of food but on foods of lower nutritional quality. This lowered food quality will accentuate the difficulty of maintaining the required energy intake of the greater body mass, meaning that malnutrition may well be more common among males than among females. This may lead directly to death, or through debilitation may increase the risk of predation.

Predation itself is of course a major potential cause of mortality for both males and females; indeed, it might normally be expected to represent the major cause of death in natural populations. Its impact may generally be underestimated, however, or rather its potential not fully realised. As a result of Man's activities, many deer populations survive today in areas where predators have been eradicated, or at least significantly reduced in density. Thus most studies of deer populations have been undertaken in areas where the densities of large carnivores are artificially low; in consequence, estimates of the impact of predation in a more 'natural' situation may be too low. Further, since many populations of deer *do* now live in environments such as these, where the density of large predators is artificially reduced, it may well be that predation is genuinely having a lesser effect on deer populations in general than it might do in an undisturbed system. One illustration of just how serious an impact predation may have in a more 'natural' system is, however, provided from Mishra's (1982) studies on deer populations of the Chitawan National Park in Nepal. In this area, Mishra estimated that for muntjac, hog deer, sambar and chital, predation accounted for virtually all mortality within each population (Table 5.3).

Table 5.3 Causes of death of deer found as carcases in Chitawan National Park from 1978 to 1981 (data from Mishra, 1982)

Species of deer	Cause of death		
	Killed by tiger	Killed by leopard	Other causes
Muntjac	1	4	0
Hog deer	23	1	1
Sambar	43	7	0
Chital	54	0	1

By contrast, and perhaps a more typical figure, Borg (1970) found that, in Sweden, out of 2,827 roe deer examined only 382 (13.5 per cent) died as a direct result of predation.

POPULATION DYNAMICS

The importance of predation as an agent of mortality is also clear from studies of population dynamics. Populations of deer in relatively complex, multi-species systems, sharing their environment with an abundance of potential competitors or predators, may show considerable fluctuation in numbers over time; in practice, however, their numbers vary between reasonably well-marked limits, and may be seen to fluctuate about some defined equilibrium point (Figure 5.8). By contrast, where deer are introduced to new environments where no predators are present, or when, within existing ranges, populations of predators are persecuted or even exterminated by Man, the characteristics of the deer populations alter dramatically.

Thus, for example, in agricultural areas of the Western USA during the early part of this century, wolves (*Lupus* spp.) were eliminated from much of the range of mule deer and black-tailed deer, and populations of cougar (*Felis concolor*) were also reduced to extremely low numbers,

Figure 5.8 Population change in white-tailed deer in the Llano Basin, of Texas from 1954 to 1981 (data from Teer, 1984). Total range area within the Basin is taken as 212,470 ha (525,000 acres)

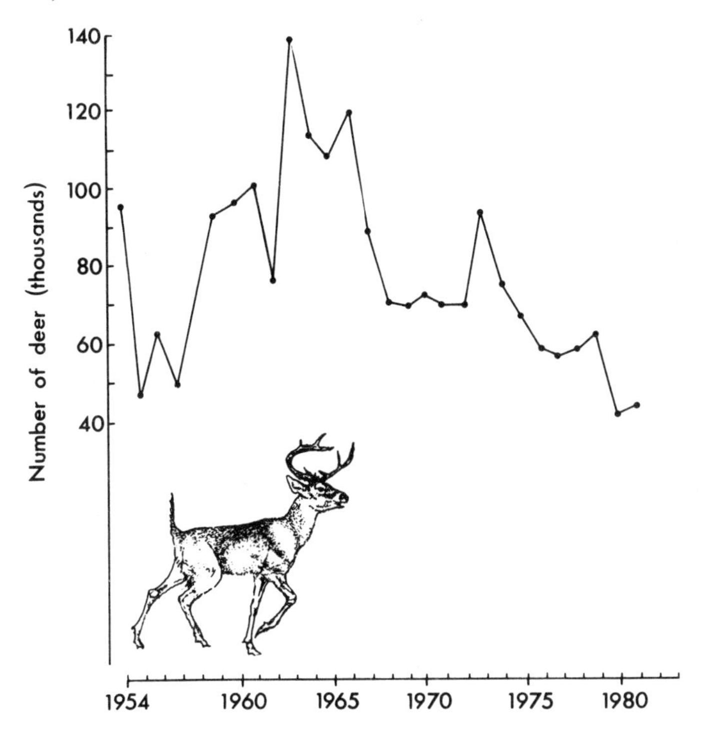

because of their depredations on domestic livestock. In response, populations of the deer themselves were observed to increase rapidly in numbers, until by the 1930s some habitats were completely saturated — to the extent that the deer populations began to experience extremely high levels of overwinter mortality (Raedeke and Taber, 1985). More extreme illustration of what may happen to deer populations in the absence of predators (and hence, by association, what may be assumed to be the normal influence of predators when they are present) derives from situations where populations of deer have been introduced to new and predator-free environments: as, for example, when red deer, sika, wapiti, fallow and white-tailed deer were each independently introduced into New Zealand (which has no native mammal fauna at all and thus offered the deer neither natural competitors nor predators), or when rusa deer were introduced into Papua New Guinea. In such situations, again, numbers increased unrestrained until the animals reached extraordinarily high densities, densities far higher than the 'normal' in their native lands (e.g. Challies, 1985; Fraser Stewart, 1981). In both these cases in fact, as with the increase in numbers of black-tailed and mule deer due to predator reduction in the USA in the 1930s, Man intervened and, through culling, 'replaced' to a degree the impact of more natural predators.

Where Man does not intervene populations may continue to expand until they far outstrip the capacity of the environment to support them. Such unchecked growth has two stark consequences. Firstly, because the

Figure 5.9 The fate of populations of reindeer introduced onto two oceanic islands in the Bering Sea (data for St Paul Island from Scheffer, 1951; for St Matthew Island from Klein, 1968)

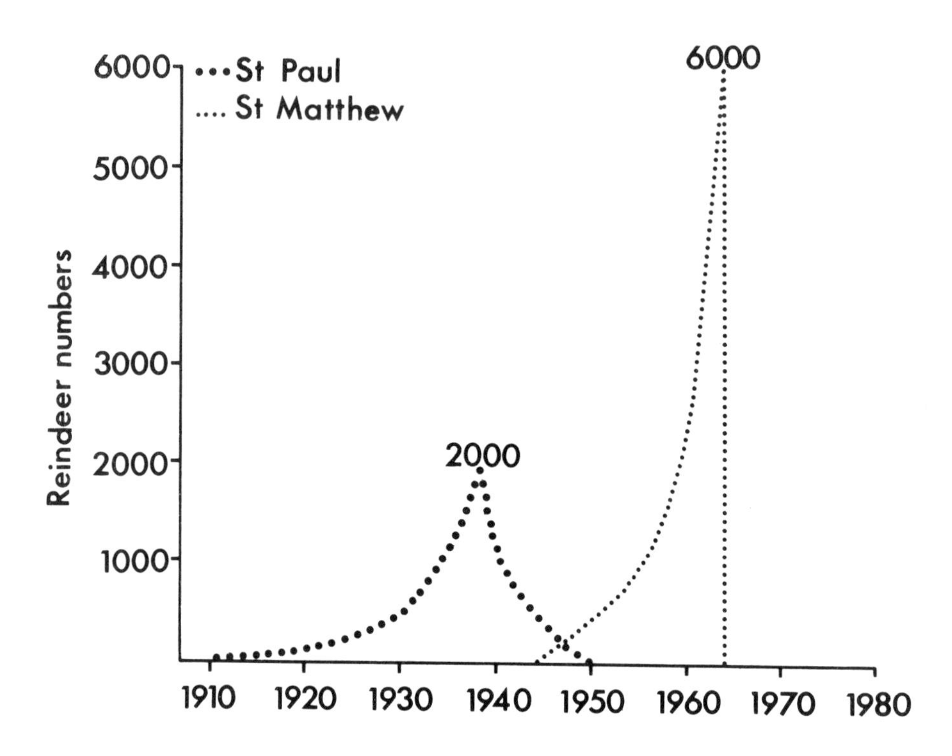

animals have exceeded the carrying capacity of the environment, they begin to cause actual damage to those very resources on which their populations depend. Secondly, because numbers are in any case far higher than the available food supply can support, there is insufficient food to go around and animals face starvation in their thousands. Figure 5.9 shows the changes in population numbers of a herd of reindeer introduced in 1944 to St Matthew Island in the Bering Sea (to provide a source of fresh meat for US coastguards stationed there). From the time of their introduction until 1963 they clearly increased extremely rapidly, perhaps at near to their maximum rate, reaching an estimated peak abundance of some 6,000 animals in the summer of 1963. It was obvious at this stage, however, that they had passed the capacity of the environment to support them; indeed, vegetation showed clear signs of overgrazing even as early as 1957. In the winter of 1963/4, with the problems of overpopulation accentuated by particularly severe weather conditions at the time, there occurred a catastrophic die-off, and the population crashed to leave, in 1964, a total population counted at 42 individuals. Similar patterns of initial rapid increase followed by a tremendous crash in numbers within 30 or 40 years are also reported in other reindeer populations introduced to other small arctic islands (Scheffer, 1951; Klein, 1968).

Figure 5.10 Population changes in reindeer introduced onto South Georgia (based on data from Leader-Williams, 1980)

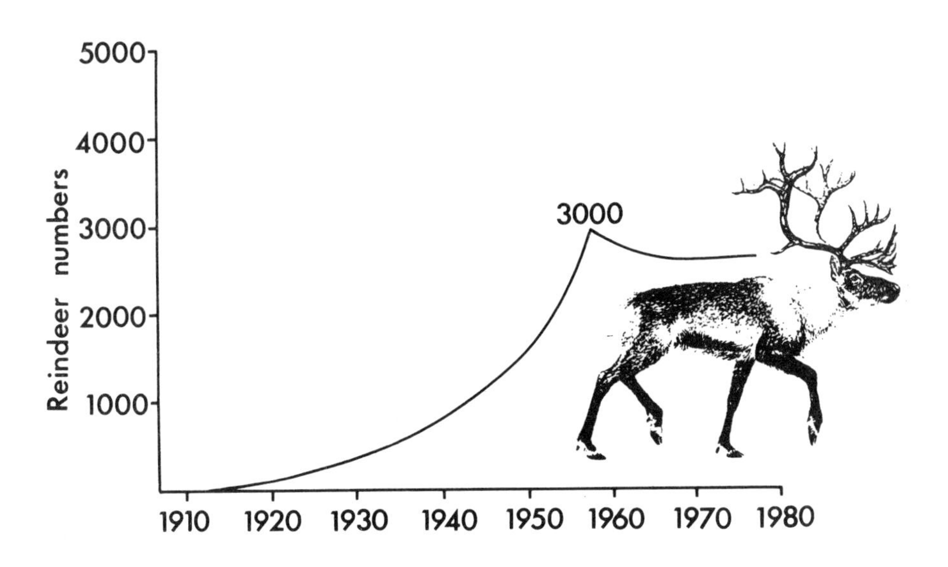

Such examples are, however, somewhat extreme. While populations of deer (and indeed all ungulates) introduced to new environments characteristically do show such a phase of irruptive growth followed by sudden decline (Riney, 1964; Caughley, 1970), they more usually do not decline quite so dramatically or quite so far. Rather, as numbers crash once more below the true carrying capacity of the environment, the rapidity of the decline more commonly slackens. Always presuming that resources have not been permanently damaged by the previous high population levels, declining animal numbers allow them to recover and regenerate: food resources increase again, the herds can feed and breed once more, and numbers rise again, this time more 'cautiously', to approach a true balance with the realistic carrying capacity of the land. Reindeer populations introduced to the antarctic island of South Georgia in 1911 and 1925 (Leader-Williams, 1980) illustrate this more typical pattern of population change (Figure 5.10).

The initial, irruptive phase of growth of populations such as these, in predator-free environments, by contrast to the rather more 'conservative' growth of populations in more complex systems, emphasises perhaps the important role that predation may normally play in regulating population numbers in deer. The fact that in the majority of cases populations do eventually stabilise at some equilibrium point even in predator-free environments suggests, however, that there must also be a series of intrinsic factors which may affect population performance in these and other deer populations, in such a way that populations will eventually stabilise on their own. The equilibrium density attained through such self-regulatory mechanisms may be far higher than that which would obtain in the presence of predators, because population size is now adjusted with respect to the carrying capacity of the environment and is not limited below that level by predation pressure; but ultimately the population will stabilise. Predation, then, may certainly act to reduce maximum population levels achieved, setting population ceilings lower than those that would be set if determined purely by the carrying capacity of the environment for the deer themselves; it may also in certain cases contribute to regulation of the population at that level, but it is not necessary for population stability.

Population Regulation

The only events which can influence the net rate of growth of any population or contribute to the population size achieved at any time are birth, death, immigration and emigration. All these factors can be seen to change with population density.

Emigration increases as density rises, although this is not normally a gradual increase but rather occurs as a threshold response. Emigration levels tend in general to remain rather low until population levels reach a certain critical threshold; above that level, high rates of emigration are suddenly observed. Thus, among the reindeer of South Georgia (Leader-Williams, 1980; Figure 5.10), emigration from the initial site of introduction was relatively low until, after 50 years, the population reached a critical density; at that point, emigration suddenly increased and new populations were established in different parts of the island. Similarly, sika deer introduced into the Kintyre peninsula in western Scotland in 1893 remained within the peninsula south of the Crinan Canal until, in 1960, population densities had built up to considerable levels; only then were the first sika observed north of the Canal at Lochgilphead (Ratcliffe, 1987).

This 'threshold effect' may have a simple explanation. The risks associated with emigration are clearly very high. An individual must leave an area with which it is familiar and strike out in search of a new locality to colonise. It may never find a suitable new locality; if it does so, that new locality may already be occupied by a population at least as dense as the one the animal has just left. Even if it does reach a new range of low population density, it is unfamiliar with the area and may therefore not be able to exploit it so efficiently. Dispersal is, at worst, as we have seen on page 103, associated with high mortality; at the least, it may result in reduced food intake and lowered reproductive success. While, if all goes well, the emigrator may end up in a utopian new range of abundant resources and minimal competition; it is, evolutionarily speaking, a considerable gamble to take. Arguably it is such a high risk option that it becomes worth the gamble only when conditions in the normal range become extremely poor: in short, when it would be even more disadvantageous to remain. Thus, emigration might be expected to be a relatively infrequent occurrence until population densities reach such a level as to alter the balance of risk in favour of migration.

Mortality is also known to change with population density. Neonatal survival of juveniles is far lower in populations of high density. While it might be supposed that this was a result of mothers having insufficient food themselves to maintain an adequate milk yield, there is also evidence — in white-tailed deer, for example — that malnourished mothers are generally less solicitous towards their fawns, failing to groom them and care for them properly (Langenau and Lerg, 1976). Lowered condition of calves entering their first winter will also result in high overwinter mortality of juveniles in high-density populations. As we have previously noted, the first winter is in any case a time of high mortality; mortality rates are notably higher still in high-density populations.

As we discussed earlier, mortality rates after this first critical year of life are generally rather low, and there is little evidence of increased mortality among mature animals in general as population density rises. In polygynous species, however, (in which male survivorship is in any case found to be lower than that of females: e.g. Table 5.2) the differential between the sexes in mortality rates is exaggerated at high density (e.g. wapiti: Anderson, 1958; Flook, 1970; red deer: Clutton-Brock *et al.*, 1982; mule deer: Robinette *et al.*, 1957; Klein and Olson, 1960; reindeer: Klein, 1968; Leader-Williams, 1980).

Most commonly, responses to population density are seen in changing rates of survival, as we have just considered, or in reproduction. As early as 1950, differences in the reproductive performance of deer in populations of different density were recorded by Cowan (1950), Chaetum and Severinghaus (1950) and Scheffer (1951). In Canadian wapiti, Cowan reported that the proportion of hinds bearing twins was 20–25 per cent on one range of particularly high quality, but only two sets of twins were seen in four years on another range that was heavily grazed. In a corralled herd of white-tailed deer in New York State, the average number of fawns born to each female was 1.9 when food was plentiful, but dropped to only 0.43 when food was scarce owing to higher population densities (Chaetum and Severinghaus, 1950). In several other cases, too, the birth rate was shown to change with population density or food availability. (Density is, of course, a relative term, and any such response is, in practice, mediated not simply in terms of numbers per absolute unit area but by relation of current density to the potential carrying capacity of the range; effective density is thus influenced both by actual animal numbers and by range quality, and will be influenced by a change in either factor.)

Strictly, nothing can be done to increase birth rate above the physiological potential of any species; animals in low-density populations will be reproducing at their maximum potential and cannot increase output beyond this maximum even if density falls still further. In effect, therefore, the contribution to population control of changes in reproductive rate is restricted to a decline in birth rate with increasing population density. This reduction in reproductive output may be achieved in three ways. First, among those species capable of multiple births, frequency of twin births declines with density. Secondly, the reduced food availability in high-density populations may result in females being unable, after the demands of breeding in one year, to regain condition sufficiently to conceive again in the following season and thus ending up breeding only in alternate years. (Indeed, there is even evidence that not only may they not reach the threshold weight normally required for ovulation, but that the threshold itself may actually be higher in high-density populations: e.g. Albon *et al.*, 1983.) Finally, as noted on page 102, while females may conceive as yearlings in areas with abundant food/low population density, in high-density populations age of first breeding may be delayed until two or even three years of age. All these factors were implicated by Morton and Chaetum in 1946 in a comparison of the performance of populations of white-tailed deer in New York State. Deer in the south occurred at lower density on range of better quality. Here 92 per cent of adult does were pregnant in any year; of these, 33 per cent had only single offspring, while 67 per cent had multiple young. In addition, 36 per cent of all yearlings were pregnant. In the north of the State, only 4 per cent of yearlings were found to be pregnant and only 78 per cent of adult does conceived; of these does, just 19 per cent had multiple births.

Such results have been confirmed again and again for white-tailed deer throughout the US, with perhaps the most comprehensive recent study being presented by Teer (1984). Similar results have been demonstrated in other species, too. Ratcliffe (1984), for example, has shown clear evidence of just the same changes in reproductive rate with population density and/or range quality in his extensive studies of red deer in Scottish conifer forests. Multiple births were everywhere uncommon, but the proportion of adult hinds pregnant in any one year varied among populations from 40 per cent to 90 per cent, and incidence of yearling pregnancies ranged from zero to over 90 per cent (Figure 5.11). Again, the

Figure 5.11 Response of reproductive performance to density or to range quality, expressed as the proportion of adult female red deer and yearlings found pregnant in a number of different Scottish forests (N represents the sample size in each case). Source: Ratcliffe (1984)

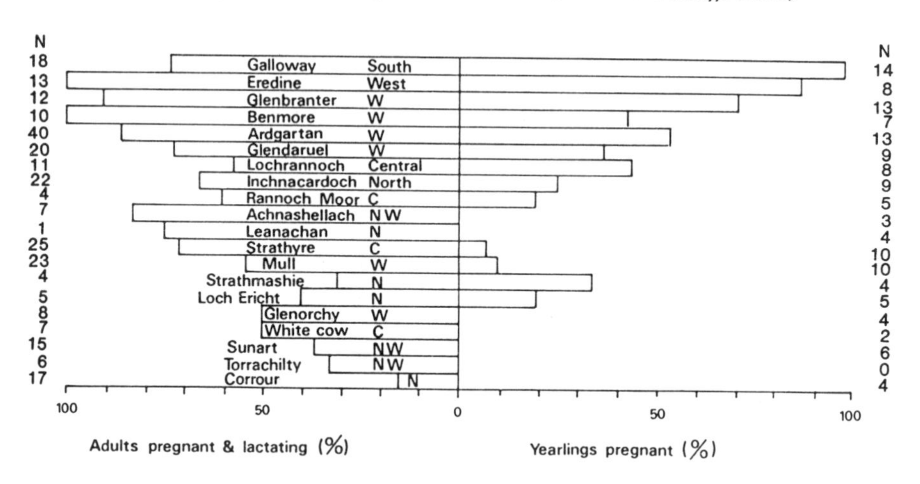

number of offspring per individual female in any one year ranged from zero to, exceptionally, three, with population means of 0.6 to 2.0 in different areas; incidence of yearling pregnancy in 'high-density populations' was low (less than 5 per cent), while in populations on productive range yearling pregnancy approached 90 per cent and animals occasionally conceived in their first rut (at perhaps three months of age: Ratcliffe, Bathurst and Putman, unpublished).

As density increases within a population, therefore, mortality rates rise (particularly among juveniles) and birth rates fall. In large part this may be due directly to changes in body condition, with animals in high-density populations experiencing increased competition for food and thus achieving lower body weights and being less well able to maintain them. Overwinter survival falls, and, since probability of conception is related to body weight, the proportion of yearlings becoming pregnant *and* the proportion of lactating adults able to conceive again in the same year declines. Food reserves for lactation also fall and thus condition of neonates may be inadequate to sustain them through the first critical year of life, particularly if severe weather conditions are experienced. There are, however, added subtleties.

An animal's performance may be affected not just by the population density currently prevailing; it may also carry forward a legacy from the density experienced in the year of its birth (which may have affected its growth rate and the final body weight achieved). Clutton-Brock and his co-workers have demonstrated, however, in the red deer of Rhum that, while population density in the year of birth might affect early survival, it did not significantly affect survival and reproductive rate as an adult (Clutton-Brock *et al.*, 1982; Albon *et al.*, 1987).

There may also be behavioural effects. Overt aggression between individuals is increased in high-density populations; frequency of aggressive interactions rises noticeably. Subordinate females of social species may be threatened or displaced far more commonly and have their normal behaviour interrupted. In the calving season this may intrude upon maternal behaviour, interrupting suckling bouts, disturbing mother and calf, and reducing the time a mother can spend caring for her offspring; neonatal mortality may increase. Equally, during the rut, competition for males in high-density populations may be such that subordinate females are kept away from attentions of the master stag of a harem or rutting stand by continually being displaced by more dominant animals. Male-male competition at the rut also increases in high-density populations, and males may spend so much time repulsing the challenges of potential rivals that they do not have sufficient time left to ensure fertilisation of all the females they are fighting to retain. All these factors may also contribute to a reduction in population growth at high densities.

Finally, we should not forget another behavioural characteristic which may also have a profound effect on population density. Many of the factors described above which may influence the size or growth rate of populations relate primarily to the more social species of deer. Yet, at a more fundamental level still, territoriality itself provides a mechanism for regulation of population density in a given area. Since territoriality implies exclusive rights to a particular area of ground, then, whether or not territory size is related in any way to availability of resources, it must set a limit to the number of animals which may be supported in a given area, and thus overall limits to population density. Indeed, we have seen how the maintenance of a territorial system results in tremendously high dispersal of juveniles from the natal range in such species (page 103). There is some evidence that, as population density increases in an area, the absolute size of territories may decline (e.g. Loudon, 1987), but clearly this contraction of territory size can continue only up to a certain point, below which the territory becomes unviable. Once that point is reached, since no new individual may then become established within the area except by displacing an existing territory-owner or occupying a territory vacated by the previous owner's death, an overall limit to population density is established.

Population densities may be limited naturally by predation, by territoriality or by density-related changes in rates of birth, death and emigration. In practice, however, there are perhaps few populations whose numbers are not now affected by Man's management. Chapter 8 will review in more detail relations between human and deer populations, but it would be inappropriate to end this section without recognition of this latter point: that whether indirectly, through habitat erosion, competition from domestic livestock, etc., or directly, through hunting or through more controlled management for conservation, exploitation or control of pests, Man himself provides perhaps the greatest influence on sizes of most deer populations and in many cases himself assumes responsibility for maintaining population numbers at the levels he has determined.

6 Interactions with other species

PREDATION

In the last chapter we noted that predation, at least in areas where numbers of large carnivores have not been artificially reduced by Man, may be a major cause of mortality among deer populations. In many cases it may be *the* major cause of mortality, and many have claimed that in natural systems predation is the primary factor limiting or controlling population size. While we showed that increasing population density has a number of effects on rates of reproduction and 'natural' mortality within the population which will ultimately themselves act to limit population growth and result in stabilisation of population size at some equilibrium level, many authors would argue that few populations in practice ever reach a size at which such intrinsic factors would take effect. They argue rather that, where densities of natural predators have not been artificially lowered, most deer populations are directly limited by predation, at levels well below that at which they would otherwise stabilise. Mishra (1982) showed that, in his study area in Chitawan National Park in Nepal, predators accounted for nearly all observed mortality in muntjac, hog deer, chital and sambar. In other species, too, predation contributes a high proportion of neonatal mortality (for even such relatively small species as lynx, caracal or coyote can have an impact on deer populations through predation on calves: Horn, 1941; McMichael, 1970; Trainer, 1975; Schladweiler, 1976; Robinette *et al.*, 1977) and may account for nearly 100 per cent of adult mortality. Predation levels are not, however, always so high. Pimlott (1967) estimated that wolves killed each year about 37 per cent of a population of white-tailed deer in Algonquin Park in Canada, and Mech's (1966) figures for predation by wolves on a population of moose in Isle Royale National Park in Michigan estimate that losses were approximately 25 per cent a year.

In effect, there are probably very few populations in which a truly natural balance may be observed. In the majority of cases, this natural balance has been disrupted by Man, who has reduced or eliminated the larger predators and, subsequently, has assumed responsibility himself for management of the deer populations. Thus it may be hard fully to substantiate just how general may have been the role of predators in regulating population numbers of the deer on which they preyed. Houston (1982) concludes that:

> 'Generalizing about the effects of predation on the abundance of large ungulates is difficult; the number of such studies is limited, they are inherently difficult to conduct, and they show that the effects of predation vary. A number of recent studies suggest, however, that predators are unable to prevent large ungulates from being resource limited in several natural ecosystems, particularly if the large ungulates are migratory or are the most abundant of several alternate prey (Hornocker, 1970; Schaller, 1972; Kruuk, 1972; Carbyn, 1974; Sinclair, 1979). The series of excellent studies of moose-wolf relationships on Isle Royale now indicate that moose are resource limited (Peterson, 1977) instead of being held below this level by predation, as suggested earlier (Mech, 1966). It would appear to be in situations where predation by native carnivores and modern Man are combined or when food for the prey is reduced by particularly severe environmental conditions that predation may actually act to reduce population numbers in these larger species.'

Houston goes on to note, however, that, in contrast to the effects of native predators on large ungulate species, the smaller, less abundant ungulates with restricted distribution may more often be limited by predation (Kruuk and Turner, 1967; Schaller, 1972).

Conversely, of course, deer themselves (as the basic food source for the predators) may have a profound effect on the carnivore populations which prey upon them, particularly when the predators are heavily dependent on only one, or relatively few, species of prey. White-tailed deer provided the primary source of food for the wolves of Algonquin Park (Pimlott, 1967), with over 80 per cent of wolf faeces collected in summer containing deer remains; the next most abundant food source, moose, comprised by comparison a mere 8 per cent of the wolves' total food supply. Under such conditions we might expect fluctuations in availability or performance of the deer herds to influence predator populations, as indeed has been demonstrated by Mech (e.g. 1970) and Mech and Karns (1977).

COMPETITION

Deer share their environment of course not only with predatory species but with a whole variety of other animals, many of which may be potential *competitors*, and, like predation, intense competition for food or other resources may have a major impact on population performance or ecology. Competition, whether direct (resulting in actual confrontation between the two competitors and direct contest for use of some shared resource) or indirect (where one animal affects another's use of the resource by using some of it up, leaving less available to the 'second-comer'), clearly interferes with an animal's access to resources and may reduce resource availability overall. If resources are limited in supply, this may have a depressive effect on population performance, restricting both the rate of population growth and the ultimate ceiling reached. Intense competition of this sort constitutes a powerful selection pressure, and ultimately may result in a change in the behaviour or ecology of one or other species, either through immediate change in the behaviour of individuals or, in the longer term, through evolutionary change within the species, but in either case shifting to use of a slightly different spectrum of resources in order to reduce the level of competition experienced.

Few deer populations exist in isolation, as the only herbivores within their range; most share their environment with a variety of other large ungulate species. For the most part, competition in the past, or the potential for it, has led to the evolution of a high degree of ecological separation, so that each species today has already become specialised to a different niche, to use of a different set of resources within the system, specifically to avoid just this kind of competition. Thus, where a number of species co-occur within a given geographical area, they will differ markedly in the habitats occupied or will be specialised to exploit different types of food (Chapter 3). The separation is not, however, complete, and where a number of species of deer do occur together, or where deer occur with other species of ungulates, just how much interaction is there?

Studies by Hanley (1984) of black-tailed deer and wapiti during the summer months in an area of forest in Washington State, USA, showed that, while the two species occupied the same basic geographic area and apparently similar gross habitat (spruce forest), they selected different patches within the forest, of different growth stages or differing species composition. In each case habitat preference proved to be consistent with the expected feeding habits of the two species: with wapiti more of a bulk feeder than the black-tailed deer. Thus, black-tailed deer showed a high preference for habitat patches dominated by browse and forb species, while wapiti showed a high preference for areas whose vegetation was dominated by grasses and forbs. In an elegant analysis, Hanley showed that the two species in effect separated out quite clearly in ecological habit and were unlikely to compete. Hanley's observations were, however, restricted to the summer months. Similar studies, by Jenkins and Wright (1988) of habitat selection by white-tailed deer, wapiti and moose in Montana, extended analysis into the winter period. Here again, at least in mild winters, the three species showed distinct preferences for different community types and separated out quite distinctly from each other with respect to the snow depth and density of vegetation preferred. White-tailed deer preferred mature coniferous forests and avoided more open communities of younger growth stages; wapiti, like those of Hanley's study, preferred open areas with a greater abundance of ground vegetation (grasses and forbs); and moose showed clear preferences for dense shrub and wetter areas.

In severe winters, however, when temperatures were low (and shelter therefore a more critical commodity) and there was deep snow covering the ground, the wapitis' behaviour changed and their selection of habitat types was very similar to that of the white-tailed deer. Jenkins and Wright in fact report a high level of overlap in habitat use between wapiti and white-tailed deer, and also between wapiti and moose during more severe winters; they conclude that overlapping distribution and habitat-use patterns, coupled with resource limitations resulting from the deep snow cover, suggest a high potential for interspecific competition in such winters between wapiti and both white-tailed deer and moose.

While a potential for competition may exist in this case, actual competition is notoriously hard to prove. Perhaps the most obvious place to look first for evidence of such competition is in New Zealand, where ten different species of deer were introduced between 1861 and 1910, of which eight have successfully established themselves in the wild. In such a situation, all the deer found themselves in an unfamiliar environment, faced with novel conditions and vegetational types of which they could have had no previous 'evolutionary experience'. More importantly, as introduced species they can have had limited prior opportunity for evolutionary change in each other's presence to establish ecological separation. The dominant species in the wild are the red deer, which is now present in most forested and high-country areas, and fallow, which have survived in 14 separate localities. The other successfully introduced deer (wapiti, moose, sambar, rusa, sika and white-tailed deer) are each confined to one or two localised populations. Challies (1985) notes that some degree of competition between red deer and several of the other species is apparent. The best-documented example is that between red and sika deer: high numbers of sika deer appear able to displace red deer in time, especially in lower-altitude forests; this supposedly results from sika deer having a closer-feeding habit, which allows them to thrive on range already depleted and less suitable for red deer (McKelvey, 1959; Kiddie, 1962). There is also circumstantial evidence for some form of competitive exclusion of red deer by both fallow and white-tailed deer (Kean, 1959). Where fallow deer have established and reached high numbers before coming in contact with red deer, they have tended to remain the dominant species on at least part of their range, especially at lower altitudes. In the Blue Mountains, where fallow deer have been present in high numbers since around 1900, red deer have not become established. White-tailed deer have thrived in the low-altitude forests of Stewart Island to the exclusion of red deer, which are now present in only low numbers in less favoured habitats (M.J. Slater, 1982, in Challies, 1985).*

Citing this particular example of New Zealand as evidence for some degree of competition in deer is perhaps a little artificial, for as noted it was deliberately selected as one likely to exaggerate the competitive interactions, since all species involved were non-native. The truth is that, in more 'natural' systems, evidence for competition is hard to find. For the most part, evolution over many generations has established the separations/specialisations that we have referred to above (and see also Chapter 3). Thus, as Mishra (1982) found in his studies in Nepal, while four species of deer (hog deer, chital, sambar and muntjac) co-exist within the Chitawan National Park, their habitat preferences and diets (pages 36–7) are so distinct that there is little likelihood of competition.

A more formal analysis of interactions within another such multi-species community of deer (in the New Forest of southern England, where

---

* Challies (1985) makes the interesting observation that, in all these examples, it appears to have been the smaller-bodied of the competing deer species that has in each case been the more successful: an observation entirely in keeping with the arguments of pages 62–3 that larger-bodied animals are generally at a disadvantage when in competition with smaller animals, of the same or different species.

red deer, roe deer, fallow and sika all occur together) also found little evidence for competition (Putman, 1986). Although on the basis of patterns of habitat use alone, or dietary composition, considerable overlap was recorded between pairs of species, when overall patterns of resource use were considered, sufficient separation between the species was apparent to suggest that they were unlikely to compete in practice.

The actual degree of overlap between any two species in terms both of habitat use and of diet may be formally examined in calculation of an 'index of niche overlap', such as that of Pianka (1973). Such indices examine objectively the extent to which a number of species overlap in their use of a particular ecological resource (food, habitat, time, etc.). We may calculate for each species what proportion of its needs in a given resource is met by different parts of that resource and then examine the extent to which this overlaps the pattern of use of those same resources by the other species (see e.g. Pianka, 1973; Putman and Wratten, 1984). The index may assume values from 0 (total ecological separation) to 1 (total overlap). Calculated for the deer of the New Forest in terms of diet and habitat use, these indices allow a more objective assessment of overlap (Figure 6.1).

For most of the year little dietary overlap is observed between roe and the other two deer species; overlap between roe and sika is consistently low, but in winter, when food is restricted in both quantity and variety, overlap between roe deer and fallow increases significantly. Jackson (1980) also noted that diets of roe and fallow deer within the New Forest showed greatest overlap in winter and early spring. Concentrating on this period of the year on the assumption that, if there is any competition for food between the two species, it is likely to be at its most intense at this time when food is shortest, he nonetheless concluded that widespread competition is unlikely to occur. Staple winter foods common to both species were coniferous browse, dwarf shrubs, fruits, bramble, rose and ivy, but there are clear distinctions in the relative importance that each food has in the total intake. By contrast, diets of fallow and sika deer show significant overlap throughout the year: both species are intermediate feeders on Hofmann's classification (page 47), and clearly both select the same types of food. With respect to their use of habitat, roe deer clearly differ from fallow and sika, but, once again, sika and fallow deer do show a considerable overlap, particularly through spring, summer and autumn.

The actual ecological separation achieved in practice between the species is, however, a cumulative effect of separation in the two resources of habitat use and diet combined, and may be derived mathematically by multiplying together the two separate indices of overlap in habitat use and diet (Figure 6.2). On theoretical grounds it has been calculated that, where overlap is restricted to a level less than 0.54, the animals are unlikely to be experiencing severe competition (MacArthur and Levins, 1967), and it is clear from the figure that, when patterns of use of all resources are taken into account in this way, overlap between the species falls below this critical level.

The New Forest is grazed not only by deer, but by free-ranging cattle and ponies. Overlap indices between these two species and between them and the Forest deer are included in Figures 6.1 and 6.2 for completeness, and it is clear that the degree of overlap in resource use between these two species, and between the domestic species and, individually, fallow and sika deer, is far higher than among the deer themselves: again suggesting that competition is likely to be significant only in artificial systems where non-native species (in this case domestic animals, sika deer) have been introduced.

Such analyses of overlap should, however, be interpreted with caution. High overlap does not necessarily imply competition; indeed, almost by definition, if high overlap is observed then the animals must be exploiting superabundant resources. By converse, low overlap should not necessarily be seen as evidence of lack of competition; it could equally be that competition for shared resources has resulted in an ecological separation, that the competing species have been forced to adopt different diets to *avoid* competitive conflict.

Figure 6.1 (a) Niche overlap among New Forest herbivores. Overlap in habitat use.

a)

	Cattle	Ponies	Fallow	Sika
		Spring (February – April)		
Cattle	*			
Ponies	0.89	*		
Fallow	0.50	0.30	*	
Sika	0.52	0.30	0.61	*
		Summer (May – July)		
Cattle	*			
Ponies	0.80	*		
Fallow	0.06	0.13	*	
Sika	0.11	0.16	0.71	*
		Autumn (August – October)		
Cattle	*			
Ponies	0.65	*		
Fallow	0.19	0.29	*	
Sika	0.16	0.29	0.69	*
		Winter (November – January)		
Cattle	*			
Ponies	0.59	*		
Fallow	0.05	0.20	*	
Sika	0.03	0.30	0.49	*
		All months combined		
Cattle	*			
Ponies	0.78	*		
Fallow	0.18	0.24	*	
Sika	0.20	0.27	0.63	*

Figure 6.1 (b) Niche overlap among New Forest herbivores. Overlap in food use.

b)

	Cattle	Ponies	Fallow	Sika	Roe
		Spring (February – April)			
Cattle	*				
Ponies	0.87	*			
Fallow	0.96	0.92	*		
Sika	0.77	0.66	0.81	*	
Roe	0.20	0.14	0.39	0.53	*
		Summer (May – July)			
Cattle	*				
Ponies	0.95	*			
Fallow	0.94	0.94	*		
Sika	0.80	0.72	0.78	*	
Roe	0.14	0.14	0.35	0.32	*
		Autumn (August – October)			
Cattle	*				
Ponies	0.96	*			
Fallow	0.86	0.88	*		
Sika	0.90	0.80	0.79	*	
Roe	0.20	0.17	0.45	0.31	*
		Winter (November – January)			
Cattle	*				
Ponies	0.93	*			
Fallow	0.65	0.63	*		
Sika	0.77	0.69	0.87	*	
Roe	0.16	0.14	0.68	0.37	*
		All months combined			
Cattle	*				
Ponies	0.96	*			
Fallow	0.92	0.91	*		
Sika	0.84	0.75	0.84	*	
Roe	0.18	0.15	0.43	0.37	*

In fact, the diets of horses, cattle, roe and fallow deer in the New Forest are much as would be expected elsewhere, suggesting that few 'adjustments' have had to be made to permit them to co-exist within the Forest — and thus implying little direct competition. In sharp contrast, the diet of New Forest sika differs markedly from that which they appear to adopt when allowed to 'do what they want' in isolation. Sika in Dorset and in five major forests in Scotland all have much the same diet, composed primarily of heather (30–40 per cent) and grasses (50–70 per cent) (Mann, 1983). Only in the New Forest does the diet seem to change, with

Figure 6.2 Combined niche overlap (all resources) among the large herbivores of the New Forest, England. Source: Putman (1986)

	Cattle	Ponies	Fallow	Sika	Roe
		Spring (February – April)			
Cattle	*				
Ponies	0.63	*			
Fallow	0.48	0.28	*		
Sika	0.40	0.20	0.49	*	
Roe	(0.20)	(0.14)	(0.39)	(0.53)	*
		Summer (May – July)			
Cattle	*				
Ponies	0.70	*			
Fallow	0.05	0.12	*		
Sika	0.09	0.12	0.55	*	
Roe	(0.14)	(0.14)	(0.35)	(0.32)	*
		Autumn (August – October)			
Cattle	*				
Ponies	0.58	*			
Fallow	0.16	0.26	*		
Sika	0.14	0.23	0.55	*	
Roe	(0.20)	(0.17)	(0.45)	(0.31)	*
		Winter (November – January)			
Cattle	*				
Ponies	0.42	*			
Fallow	0.03	0.12	*		
Sika	0.02	0.21	0.43	*	
Roe	(0.16)	(0.14)	(0.68)	(0.37)	*
		All months combined			
Cattle	*				
Ponies	0.70	*			
Fallow	0.17	0.22	*		
Sika	0.17	0.20	0.53	*	
Roe	(0.18)	(0.15)	(0.43)	(0.37)	*

increased intake of browse and lower reliance on grasses. Nor is the direction of the change what one might expect in terms of the difference in habitat: the sika deer of Dorset and Scotland are animals of coniferous plantation and heathland; the New Forest offers a wider diversity of vegetation types with, in principle, better opportunities for grazing. Such an unexpected shift in diet may well, then, be the result of competition, and it is suggested (Putman, 1986) that in the New Forest there may indeed be real competition for forage between sika and the Forest ponies.

Figure 6.3 Comparison of food use by mule deer, pronghorn antelope, horses, cattle and sheep in the Great Basin rangelands of California and Nevada, USA. The scatter-plots show the degree of segregation between the species in terms of diet in spring, summer, autumn and winter (asterisks indicate the centroid or 'mean' for each species). 1 = horses; 2 = cattle; 3 = sheep; 4 = pronghorn; 5 = mule deer. Source: Hanley and Hanley (1982)

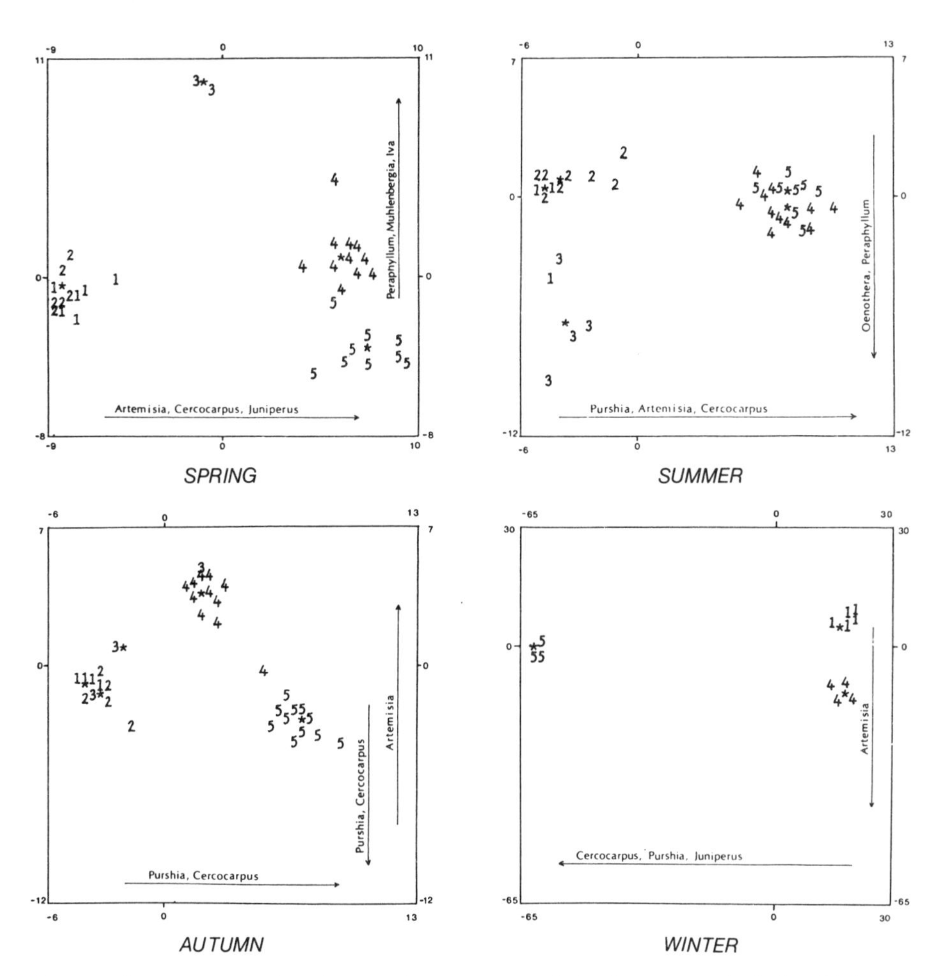

Scatter-plots are produced by the statistical procedure of discriminant analysis, by which populations (in this case of particular species) are separated out from other populations with regard to the degree of difference between the two in some particular character common to both (in this case amount of use of particular foodstuffs). The analysis selects those characters which maximise discrimination between the groups, and uses them, separately or in combination, to produce two, right-angular coordinates on which the positions of the different populations may be represented. Separation between groups (species) is thus always maximised; the characters contributing most to this segregation (i.e. foodstuffs for which there is the greatest difference in use between species) are identified beside the axes along which they contribute most to segregation. Thus, by way of example, during spring, horses and cattle (1 and 2), while very similar to each other in diet, separate out from pronghorn and mule deer (4 and 5) in amount of use made of Artemisia, Cercocarpus *and* Juniperus. *Horses differ from cattle, inasmuch as they show any difference at all, with respect to use of* Peraphyllum, Muhlenbergia *and* Iva *plants, and these same species are responsible for separation between pronghorn and mule deer. Sheep (3) are clearly different from all the other species, and this separation is due to differences along both axes. Overall, the absolute difference between the species in terms of diet is represented by the linear distance between the group centroids (*) on the scatter-plot.*

As this last example makes clear, deer do not share their environment just with other species of deer. In many multi-species grazing systems deer are found in conjunction with antelope, sheep, goats or, as in this case, free-ranging domestic stock (e.g. Hansen and Reid, 1975; Hansen *et al.*, 1977; Hanley and Hanley, 1982; Gallini, 1984). Here too, however, ecological separation is more the rule than the exception: in the Hanleys' most elegant study of food-resource use among feral horses, cattle, sheep, mule deer and pronghorn antelope, it was again only the domestic species that showed potential for competition. As in Putman's New Forest studies, the diets of cattle and horses showed striking similarity in all seasons, and closely overlapped that of the domestic sheep in autumn and winter. In all seasons the diets of the mule deer and pronghorn, however, were quite distinct, both from each other and, as wild species, from those of the domestic animals with which they share their range (Figure 6.3).

PARASITISM

Few environments in nature remain uncolonised for long and, even when the environment is itself a living organism, it is not long before some other organism will have evolved to exploit it. Parasitism is among the most successful of all modes of life, and deer, like other organisms, carry a full complement of both internal and external parasites. All species are susceptible to a number of external parasites, which live in the animals' coats and feed on blood (as do ticks, biting lice, sucking lice) or burrow into the skin (as do the larvae of warble-flies or nasal bot-flies). They may also be hosts to a wide range of internal parasites. These latter fall, broadly speaking, into three main groups: roundworms (or nematodes) tapeworms (cestodes) and flukes. Most live in the alimentary canal and each type specialises in a particular region of the gut; but there are also species in the liver (various species of liver fluke), bladder and lungs (lungworms such as *Dictyocaulus*). Different deer species are of course susceptible to different parasite species, and many of the parasites are extremely host-specific. By way of illustration, to show the type of burden which may be supported by any one species, Table 6.1 lists the various parasites which may be found in fallow deer in Britain.

Under normal circumstances, parasitic infestation will not directly kill the host (indeed, unless it has extremely efficient methods of getting from one host to another, it is a poorly adapted parasite that does so, for it kills its own life-support system at the same time). Parasites can, however, cause considerable debilitation and loss of condition, rendering the host more susceptible to infection by disease, predation etc., and it is therefore possible that, while they would not normally be considered a major direct cause of mortality, they may indirectly increase risks of mortality from other causes (page 106). In actual practice, although heavy burdens of both internal and external parasites may be observed in animals already weakened from other disease or from starvation, this is unusual and, more typically, levels of infestation in any one individual are rarely sufficiently severe as to be likely to cause serious debilitation.

Indeed, both in terms of incidence in the population (as number of deer infected) and of individual levels of infestation (numbers of parasites

Table 6.1 Parasite burdens of adult deer: data on parasite load and incidence of other diseases or abnormalities recorded in fallow deer in Britain

Pathogen	Incidence	Notes
INTESTINAL PARASITES		
Roundworms (Nematoda)	81–100%	58 species recorded in fallow deer; numbers of worms usually lower than found in domestic stock
Tapeworms (Cestoda)	c.12%	6 species; extent of infestation varies markedly with area
Flukes (Trematoda)	Variable	4 species affect fallow deer. Degree of infestation varies from area to area: rare in e.g. Epping Forest (Chapman and Chapman, 1975), while almost all New Forest fallow show signs of present or past attack (McDiarmid, 1969)
OTHER INTERNAL PARASITES		
Lungworm	Unknown	2 species; may lead to parasitic pneumonia; major cause of overwinter mortality in other species (e.g. roe deer)
EXTERNAL PARASITES		
Biting lice (Mallophaga)	20–70%	*Damalinia tibialis*. Infestation seasonal: highest late spring, early summer.
Sucking lice (Siphunculata)	Recorded	(*Solenopotes burmeisteri*)
Ticks (Acarina)	Up to 100%	3 species occur on fallow deer, but primarily *Ixodes ricinus*; widespread, though level of infestation varies from place to place
Mites (Acarina)	Recorded	Causes *Sarcoptes* mange
Flies (Diptera):		
Warble- and bot-flies	'common' in Europe; not regularly recorded in UK	—
Keds	Up to 100%	*Lipoptena cervi*, bloodsucking fly of family Hippoboscidae. Seasonal in incidence: peak infestations late autumn/winter
OTHER ABNORMALITIES		
Ring bone	50%	Data from Chapman and Chapman (1975)
Jaw abscess associated with weakening of bone	'Common'	Data from Chapman and Chapman (1975)
Deformity of metapodials (distal limb bones)	Up to 60%	Data from Chapman and Chapman (1975)

Data in the table derive from various sources: notably McDiarmid (1969), Chapman and Chapman (1975), Chapman and Putman (1988).

supported by those individuals which are affected), parasite loads in deer are considerably lower than those encountered among domestic stock. This difference may be due in part to the fact that wild deer usually occur at far lower densities than do artificially maintained domestic stock and range over much greater areas (thus being less exposed to infection than domestic animals pastured at high density on limited range), but it seems clear that this is not the only reason. In what is now a classic study of the parasites of red deer, sika deer, fallow deer and muntjac together with domestic horses and sheep in Knebworth Park, Hertfordshire, in England (where deer, sheep and horses all shared the same pasture, were thus at equivalent density and were exposed to the same 'challenge' from parasites), Mary Holt showed that worm burdens among the deer never reached the levels recorded for the domestic animals; indeed, although faecal egg counts showed both sheep and horses to be infested and with what might be regarded as 'normal' parasite burdens, helminth eggs were found in faecal samples from only one individual red deer, and in quantities well below those usually regarded by veterinarians as representing clinical infection (Holt, 1976). Since exposure to parasitic infection was the same for sheep and deer, such results suggest that the susceptibility of deer to parasitic infestation is somehow lower than that of the domestic animals.

Holt's study was repeated and extended in 1987 by Dawn Hawkins, who revisited Knebworth and re-assessed parasite burdens of sheep, red deer and sika deer within the Park. Faecal samples of wild muntjac and fallow deer beyond the Park boundary were also examined, since, while not now maintained within the Park itself, individuals of both species could jump in and out of the area relatively easily. Hawkins assessed parasitic burdens of all species in summer, autumn and winter. In this study measurable levels of parasitic infestation were recorded in all deer species in one season or another, but in terms both of the number of different parasite species supported and of overall burden (as total number of eggs recorded per gram of faeces) levels were once more far lower than those recorded for the sheep (Table 6.2). Among the deer themselves, Hawkins's results reveal a significant overlap in the pattern of infestation in the congeneric red and sika deer, which had a significant number of species in common in all seasons. Parasites of muntjac and fallow differed in all seasons from those of the emparked species and infestation levels were consistently lower than those of either *Cervus* species.

OTHER ANIMAL INTERACTIONS

Not all relationships between deer and other animal species need be so negative. Parasitism, predation and competition are all situations in which the deer suffer, whether potentially or actually, from the interaction; but there are a few instances of mutualism, where the partners in such interaction both benefit. Among mammals, such mutualistic interactions are not common. One of the best known is that occurring between baboons (*Papio anubis*) and impala (*Aepyceros melampus*) in Africa. Here at the woodland edge, mixed feeding groups of baboons and impala are commonly encountered. It is believed that both species benefit from the association: the baboons, ever alert and with

Table 6.2 Parasitic infestation of deer and sheep at Knebworth Park, Hertfordshire, England, 1986/7 (data from Hawkins, 1987)

Host	Season	Parasites recorded	Proportion of individuals infected	Total number of eggs per gram of faecal matter in infected individuals
Red deer	Summer	Ovian ostertagian	3/14	15
	Autumn	Ovian ostertagian	2/15	21
		Chabertia sp.	4/15	
		Cooperia sp.	4/15	
	Winter	Ovian ostertagian	6/17	9
		Chabertia sp.	1/17	
		Capillaria sp.	1/17	
Sika deer	Summer	None	0/8	0
	Autumn	Ovian ostertagian	2/9	29
		Chabertia sp.	2/9	
		Cooperia sp.	3/9	
		Moniezia sp.	2/9	
	Winter	Ovian ostertagian	2/11	27
		Chabertia sp.	2/11	
		Moniezia sp.	6/11	
Muntjac deer	Summer	*Oesphagostomum* sp.	2/12	8
	Autumn	*Trichostrongylus* sp.	1/9	5
	Winter	*Nematodirus battus*	1/4	3
Fallow deer	Summer	None	0/5	0
	Autumn	None	0/4	4
	Winter	*Oesphagostomum* sp.	2/4	25
Sheep	Summer	Ovian ostertagian	1/12	433
		Moniezia sp.	5/12	
		Nematodirus battus	1/12	
	Autumn	Ovian ostertagian	9/16	228
		Chabertia sp.	9/16	
		Cooperia sp.	4/16	
		Capillaria sp.	2/16	
		Moniezia sp.	4/16	
		Nematodirus battus	5/16	
		Oesphagostomum sp.	1/16	
		Trichostrongylus sp.	2/16	
	Winter	Ovian ostertagian	4/18	53
		Chabertia sp.	1/18	
		Cooperia sp.	1/18	
		Moniezia sp.	7/18	
		Nematodirus battus	1/18	
		Strongyloides sp.	1/18	

keen senses, respond quickly to potential predators at a considerable distance, and thus, through their reactions offer the impala, too, an advance early-warning system of possible predation; the impala, for their part, moving through the grass and scrub, feeding on the vegetation, disturb insects and other potential food items for the baboons. Newton (1984) has reported a similar relationship between chital and grey langurs (*Presbytis entellus*) in areas of India. Here, however, the roles are reversed: it is the alert chital that sense possible danger and through their own behaviour warn the langurs of the proximity of tiger or leopard; and it is the chital that gain from the foraging activities of the monkeys. Grey langurs are leaf-feeding monkeys which forage in the trees above the reach of chital; but they are selective feeders and reject much of the foliage that they pluck — particularly leaf blades, which they drop after eating the stalks. Newton has estimated that a troupe of 20 langurs drops some 1.5 tonnes of foliage each year, of which 0.8 tonnes is suitable forage for chital. This extra food is probably particularly valuable to the chital in the dry season (November–June), when grass is sparse and sere. Newton notes that during the monsoon season, when other food is more plentiful, the number of times langurs and chital are observed together declines dramatically.

EFFECTS OF GRAZING ON THE VEGETATION

Before leaving this consideration of interactions between deer and other species with which they share their world, it is appropriate perhaps to consider that deer, as plant-eaters, must affect the very plants and plant communities on which they feed.

Although, typically, herbivores remove perhaps only in the region of 10 per cent of the annual green-matter production in any community, in certain instances their impact may be far in excess of this. Wiegert and Evans (1967) estimate that ungulates may remove between 30 per cent and 60 per cent of primary production of East African grasslands, and Sinclair and Norton-Griffiths (1979) calculate that herbivores (both vertebrate and invertebrate) in the Serengeti National Park in Tanzania are removing up to 40 per cent of the annual primary production. Nor is the impact accurately measured by the amount of material actually ingested. Animal feeding is often wasteful: red and roe deer browsing in deciduous forests in southern Poland remove some 46 kg (dry weight) of browse material per hectare, of which they actually consume only 19 kg (Bobek *et al.*, 1979); the balance is destroyed during feeding.

Further, it is clear that grazing may have a very much greater impact on the community than is suggested by mere consideration of the absolute quantities of plant material removed: a herbivore ingesting perhaps 10 per cent of a plant's production is going to have far more significant effect upon the plant if that 10 per cent is made up of primordia, destined for future growth, than if it merely results in a loss of 10 per cent in the form of mature leaves. In the Polish study, potential browse production of an unbrowsed forest was estimated as 172 kg/ha, though actual production in practice (including the 46 kg later removed by deer) totalled only 160 kg/ha; damage by deer suppressed productivity by 12 kg per hectare, in *addition* to its effect in removing 46 kg of that production (Bobek *et al.*, 1979). In another, similar study, browsing by moose in a pine/mountain-ash forest in Russia reduced forage biomass from 181 kg to 109 kg per hectare: only 3.5 kg of this loss was directly related to moose feeding; the remaining 68.5 kg were lost as a result of the reduced growth rate of the damaged trees (Dinesman, 1967). In fact we may note a whole variety of these more subtle effects of grazers upon vegetation, by which quite minimal absolute consumption by herbivores may have far-reaching significance.

We may see a change in productivity. We have noted here a suppression of plant production owing to herbivore pressure, but under other circumstances grazing may equally result in an increase in productivity. The effects of herbivores upon vegetational production in this way may be direct, through defoliation in feeding, or indirect: the trampling effects of hoofed mammals may lead to soil compaction and thus affect plant growth, while return of dung and urine to the system, resulting in local changes in soil-nutrient status, may also affect productivity. Direct removal of plant tissue by deer and other herbivores can affect rate of photosynthesis, respiration rate, location of nutrient storage, growth rates and phenology of the affected plant.

While, as we have noted, heavy grazing and browsing may reduce production, moderate levels of grazing may actually increase plant productivity, through stimulating some compensatory growth. Growth will be inhibited, as in our example of deer browsing in Polish forests, where herbivores damage the growth primordia of the plant, or where excessive defoliation reduces the effective leaf area of the plant below a minimum threshold for efficient photosynthesis; but there are numerous examples in the literature where lighter grazing pressure can be shown to increase productivity (e.g. Ellison 1960; Grant and Hunter, 1966; Krefting *et al.*, 1966; Vickery, 1972; Wolff, 1978; Coupland 1979). Note, however, that each plant species reacts differently to different grazing pressure, and also to different grazers; and clearly the actual result observed in any one instance will depend both on the degree of defoliation and also on the timing at which damage occurs in relation to the growth stage and growth characteristics of the particular plant. Different plant species will have markedly different responses depending on structure and growth pattern: the level of offtake at which productivity of woody species (with terminal growth points) begins to decline is far lower than that at which production of many grasses (which grow from the base) would be suppressed. Finally, it has even been suggested that one potential stimulatory effect of grazing ungulates upon productivity may arise out of plant-growth-promoting agents which have been found in ruminant saliva (Vittoria and Rendina 1960; Reardon *et al.*, 1972). Direct growth stimulations of up to 50 per cent above control levels have been recorded following addition of ungulate saliva to surfaces of manually clipped leaves (although this effect has never been demonstrated specifically for deer, work done having involved the use of saliva of bison or giraffe).

Where defoliation is sufficient to depress productivity, continued grazing may ultimately result in eradication of particular plant species from the community. Thus, through its effect upon the individual plant, grazing can alter the entire species composition of the community: with continued eradication of species sensitive to grazing, and an increase in abundance of those species which, through chemical or physical defence, or because of their growth form, may have greater abilities to resist, tolerate or escape defoliation. Even without such deletion of species, however, defoliation may still have a profound effect on community composition. By reducing the leaf area of preferred forage species, the herbivore may reduce competition for light and space experienced by other plants, which may therefore be able to colonise or increase in abundance within communities from which they would normally be outcompeted. And there are many examples where grazing or browsing by deer and other large herbivores has resulted in marked changes in plant species composition (Chadwick, 1960; Nicholson *et al.*, 1970; Gray and Scott, 1977; McNaughton, 1979). Perhaps the oldest and most-quoted example of how grazers can affect the species composition of whole vegetational communities is seen in the development of the specific plant assembly we now associate with chalk grassland in Britain, under the influence of, first, sheep and, later, rabbit grazing (Tansley, 1922; Tansley and Adamson, 1925; Hope Simpson, 1940).

Grazing herbivores not only affect productivity and species composition of the vegetational communities on which they graze; they may also have considerable impact on nutrient cycling (which may in itself have yet further repercussions for species composition and productivity of the vegetation: McNaughton, 1979; Crawley, 1983). By feeding on plant materials, animals lock up within their own tissues essential nutrients, making them unavailable to the next 'generation' of plants. In systems where nutrients are relatively abundant and there exists in the soil a relatively large pool of 'free' nutrients, this has but little effect upon plant growth, but in other, nutrient-poor systems the effects of having significant quantities of a limited nutrient supply bound up in animal tissue may well be quite marked. While such effects are really important only in nutrient-poor systems, *patterns* of nutrient cycling may be affected by grazing in any system. Animals which feed over a wide area but defaecate in a small area can have a substantial effect on local nutrient distribution. Sheep, for example, graze widely over pasture during daylight, but congregate in camps at night or for shade; as a consequence, 35 per cent of their faeces are deposited on less than 5 per cent of the grazing area, resulting in a gradual impoverishment of the wider grazing range, but continued enrichment of small areas within it (nutrient 'dislocation': Spedding, 1971). Other animals may show different habitat preferences for grazing and for elimination, so that nutrients may be removed from some habitats and returned to others; this 'translocation' of nutrients, too, has a profound effect on the nutrient dynamics of any community.

Finally, although most work has concentrated on the direct effect of grazers upon the vegetation itself, it is clear that the influence of the herbivores is not restricted to this level, but has a whole series of knock-on effects. All the various effects of grazing and elimination result in modification of the habitat itself, and of the environment it offers to others. A changing microclimate, through structural modification of the primary vegetation, will have an effect on the secondary plant species, and animals, which may colonise the 'new' environment. Changing species composition, and species dominance, will affect relative availability of food to other smaller herbivores. In short, through the changes that the grazing process causes within the vegetation, in structure, species composition and productivity, it has at once additional effects on the rest of the community dependent on that vegetation (e.g. Putman, 1986).

In conclusion, the potential effects of heavy grazing upon vegetation may be summarised as follows: increase or decrease in primary production; alteration to nutrient cycling and gross distribution of nutrients; changes in species composition (due in turn to the selective elimination of sensitive species, changes in the competitive equilibrium of the plant community under grazing pressure and to modification of nutrient flows within the system); and changes in the actual physical structure of the vegetation itself, affecting microhabitats offered to dependent plant and animal species. All these effects are somewhat general, and will result from grazing by whatever herbivore, be it deer, rabbit or antelope; in their extreme, however, they may be quite marked and, where deer populations have reached unusually high densities, such changes in the vegetation are quite commonly recorded. The changes are, however, predictable: particular species of herbivore have specific effects, dependent on feeding style, selectivity, mouthpart shape, etc., and different effects at different densities. If these effects are understood, then the tables may be turned and Man may use the grazers in management schemes to achieve a desired change in some vegetation community. Such schemes are still tentative, but communities of red deer, fallow and roe are now being managed in conservation areas within the Netherlands, and elsewhere in Europe, in order to achieve quite specific vegetational aims (e.g. Thalen, 1984; de Bie *et al.*, 1988).

7 Antlers

For many people, the major fascination of deer lies in their antlers, and it is of course the possession of antlers that distinguishes the more advanced deer from all other groups of ungulates. The fascination, to naturalists and deer-hunters alike, derives in large part from the size, complexity and sheer magnificence of the antlers themselves. While bovids produce fine horns, rapier-straight, twisted like barley sugar or coiled into a spiral, such horns are unbranched and simple. By contrast, antlers of most species of deer are complex, branched structures, often of immense size (with the antlers of a bull moose weighing up to 30 kg and those of the extinct Irish elk up to 45 kg). Yet, again by contrast to the horns of bovids, these tremendous ornaments of bone are not retained as permanent structures, but are cast and regrown each year: a curious, and seemingly extraordinarily wasteful practice.

Antlers are in essence simple, or more usually branched, structures of bone developed on the top of the head in most species of deer (excepting only musk deer and Chinese water deer). They are usually borne only by the male; only in one species, the reindeer, are antlers borne by both sexes. (Very occasionally antlered females are recorded in other species, notably roe, white-tailed deer and mule deer, but these are very rare occurrences indeed and such females are usually sterile.) In most species the antlers first show in the second year of life* and are then cast and regrown each year, with each new set of antlers generally more complex than the last — at least until the animal reaches over-maturity. This cycle is under hormonal control and is closely related to the sexual cycle (see below).

THE CYCLE OF ANTLER GROWTH

At birth, the skulls of male deer are in practice little different from those of females, but during the first year of life, although in general no external antlers will develop,* special bony growths develop from the

* In some species of deer, the first antlers are grown by fawns instead of yearlings. Thus, in moose, roe, white-tailed and mule deer, tiny antlers may be produced during the fawn's first autumn; and reindeer and caribou calves begin to sprout their first antlers within weeks of being born.

Figure 7.1 In those species which bear antlers as adults, pedicles develop on the skull of the male during the first year of life

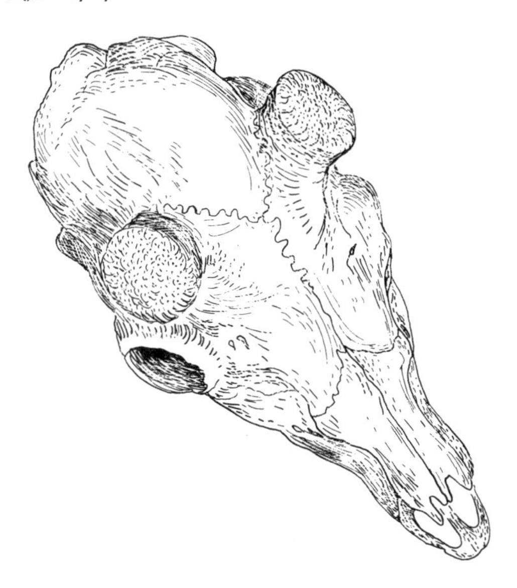

frontal bones. These 'pedicles' are essentially bony stalks growing upwards from the skull on each side of the head, terminating as a circular platform or cushion just below the skin (Figure 7.1.). It is these cushions that will support the external antler in due course (and it is at the junction with the pedicle that the shear line develops in casting of old antlers). In muntjac, the pedicle is particularly noticeable because it is extended downwards on either side of the face, giving these deer a prominent and obvious ridge on the face itself (see Figure 2.1, and Plate 3).

On the cushion provided by the pedicle, the antler later develops. The first antler is generally very simple and may be little more than a simple button or short spike. When mature, as with all antlers, the bone of the antler itself is exposed (and is actually non-living); during growth, it is covered by a furry external skin or 'velvet' (page 11) which both protects and nourishes the developing antler. The velvet is richly supplied with blood vessels and nerve endings; additional blood vessels run through special channels within the soft bone of the developing antler itself. When growth of the antler is complete, the blood supply to the velvet is cut off at the level of the pedicle; the skin withers and dries, and is rubbed off, leaving the polished bone of the mature antler exposed. After about eight months the antlers are shed, shearing off at the junction with the pedicle; the two antlers are cast usually within a day or two of each other. The skin heals above the pedicle and almost immediately the bump is apparent of the beginnings of development of the next antlers.* Except in

* Once again, many of the Odocoilinae provide exceptions to this more general rule. The casting of antlers takes place in the spring in most temperate species of deer, but in autumn or winter in mule deer, white-tailed deer, moose, reindeer and caribou. In these latter species, there may be a hiatus of many months between loss of old antlers and the onset of renewed growth the following spring.

muntjac and a few other species (e.g. pudu), in which the antler remains simple throughout life, each successive antler becomes bigger, more complex and more fully branched than that of the previous season (Figure 7.2). Antlers become progressively more complex and the main beam wider until the animal becomes senescent, when the number of branches begins to decrease (though not in any regular pattern) and the animal is said to be 'going-back'. At the ultimate, in extreme old age, the deer may once more produce no more than simple buttons; these, however, retain the width of beam of an adult animal and are thus much broader than those of yearling males.

Figure 7.2 Antler size and complexity increase from year to year as an individual grows. The figure shows nine successive antlers of one red deer stag, from the first (four-point) head

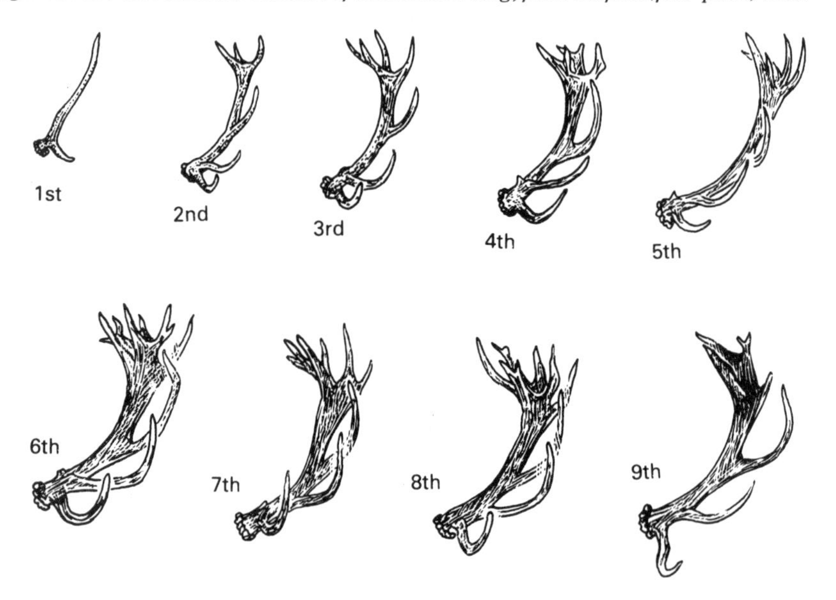

Antlers grow from the base, and the main beam of the antler develops first. Major branches may appear early as 'lumps' on the beam itself, but the full development of the finer branches and tines is completed last during the development process (Figure 7.3). Growth is incredibly rapid: the new antlers are grown within a space of only perhaps 12 to 16 weeks; yet during this time the deer must produce and mature, depending on the species, up to 30 kg of bone. It is clearly a tremendously expensive process, in terms both of the energy cost of this rapid growth and of the mineral demand it represents. In 16 short weeks the animal must effectively double the mineral deposits of its body. (Muir *et al.* (1985) calculate for red deer that a stag producing 3 kg of hard antler (of which half the material will be mineralised) will deposit 536 g of calcium and 1.0 kg of other minerals during the period of growth; over the final ten weeks of most rapid growth, the rate of calcium deposition would rise to 5 g per day.) In the search for these minerals, the deer often gnaw and nibble at their own cast antlers of the previous year, or deliberately ingest quantities of mineral-rich soil to supplement the mineral intake provided in the vegetation itself.

Figure 7.3 The annual cycle of antler growth

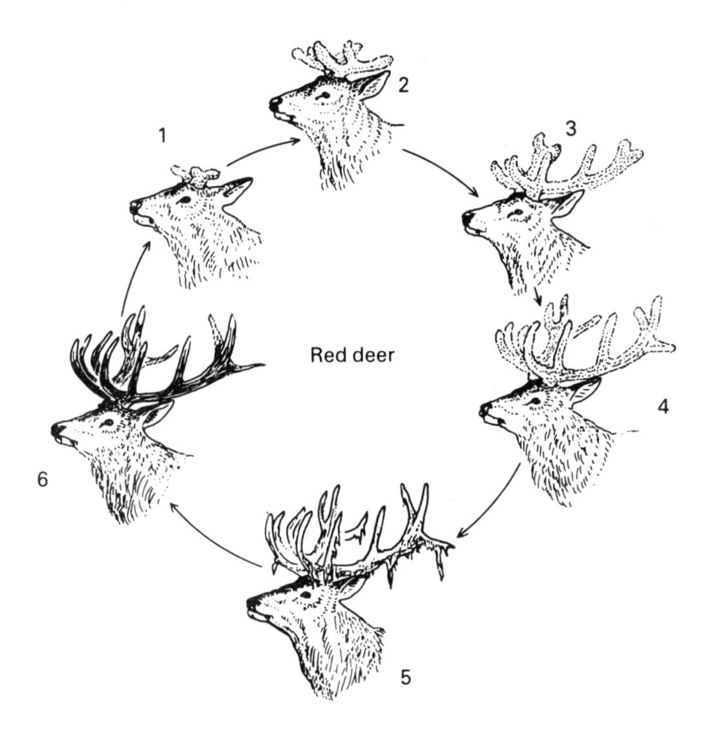

CONTROL OF ANTLER GROWTH

There is clearly great variation among different deer species in the timing and nature of antler growth. Thus, as we have noted, in roe deer, white-tailed deer, mule deer and moose, antlers as well as pedicles may develop during a fawn's first year; in the last three species, too, antlers are cast in winter and there is a clear gap between casting and regrowth not found in other temperate species. Reindeer and caribou differ again from all others in that not only do both males and females bear antlers, but that these begin to develop extremely early, in the first few weeks of life. Obviously, general principles are hard to draw. Leaving aside for the moment the reindeer and caribou, however, in all other species it is clear that the pedicle, whether it supports its first external antler during the first or second year of life, is a secondary sexual character in the strictest possible sense in that its development is associated with the onset of puberty (Lincoln, 1971; Fennessy and Suttie, 1985). This is apparent both from castration studies and from hormone studies; thus prepubertal castration will prevent pedicles ever forming, while administration of the male hormone testosterone restores pedicle development in such castrated males. Further, pedicle initiation is characteristically associated with a threshold body weight (approximately 40–50 kg in red deer stags: Fennessy and Suttie, 1985). In the same way that onset of puberty in females, in terms of the timing of first ovulation, was associated more with the time taken to reach a critical threshold weight rather than actual calendar age, so production of testosterone by males is also dependent on weight. The fact that pedicle initiation is also weight-dependent is another factor linking pedicle development to puberty.

Incidentally, the artificial administration of testosterone promotes pedicle development not only in experimentally castrated males, but also in intact females (Wislocki *et al.*, 1947; Jaczewski, 1976; Fletcher, 1978; Lincoln, 1984): showing that females of all species, not just reindeer, have the *potential* to produce antlers, but do not do so because under normal circumstances their production of testosterone is insufficient to stimulate such development. Only in exceptional cases, where the animals suffer perhaps a hormonal imbalance, will individual females produce antlers. As noted above, such females are generally, but not inevitably, sterile.

Growth of the first antler commences when the permanent bony pedicle reaches a critical length (5–6 cm in red deer: Fennessy and Suttie, 1985), and such growth would also appear to be under hormonal control, possibly by the hormone prolactin (Fennessy and Suttie, 1985). The factors which ultimately result in constriction of the blood supply to the growing bone and velvet, thus bringing growth to an end, are less clearly understood. Towards the latter stages of growth, when the antler is nearing completion, a thickened ring of bone called the 'coronet' or 'burr' develops around the base; solidification of the antler shaft itself also results in a progressive narrowing of the channels within the body of the antler in which the main blood vessels run. As a result of both these changes, there is a progressive restriction on the flow of blood through the core of the antler and to the velvet, eventually choking it off completely. The velvet dies, splits and is peeled off in shreds, revealing the now dead, bony antlers.

This clean, hard antler is essentially a dead bone attached to living bone (the pedicle); as such, it is almost axiomatic that it will be cast eventually (Fennessy and Suttie, 1985). Indeed, the more remarkable feature of the whole system is that it is actually retained for so long. Experimental castration of male deer in hard antler results in antler loss; hormone implants of testosterone maintain the antler-pedicle junction and prevent casting (Wislocki *et al.*, 1947; Lincoln, 1984). It is thus clear that the heightened levels of testosterone in adult males during the breeding season play an important role in retaining the antlers and preventing casting. When testosterone levels fall below a critical point, the antlers are cast. (The important role played by changing levels of testosterone in the antler cycle may also help to explain why, in many species of deer, older males, presumably more sexually active, tend to cast their antlers earlier than younger males, whose testosterone levels may be subject to less wide fluctuation.)

The process of casting is rapid. Within a few days, an antler, solidly anchored to the pedicle, will become detached under its own weight. The two antlers are generally cast on the same day, sometimes within hours or minutes of each other, and the changes leading up to shedding would appear to be sudden. Whitehead (1972) records an observation of two red deer stags seen fighting in a deer park only about a week or so before shedding, with such fury that, had the antlers been loosening, one or both would undoubtedly have been displaced. On another occasion a fallow buck, in trying to disengage his antlers from some obstacle, broke one of the pedicles completely off; a few days later the other antler was shed in the normal way, thus proving that prior to casting the antlers must loosen very suddenly and not by degrees (otherwise the damaged antler would have broken away from the pedicle at the coronet, instead of tearing away the pedicle itself).

Growth of later antlers follows a similar pattern to that described here for the first antler, influenced by the same hormonal changes. The initiation of growth of the replacement antler is usually thought to be a response to the wounding caused by loss of the previous antler. Thus, except for the original antlers grown by the fawn or yearling, there can be no renewed growth in the absence of wound-healing (Goss, 1985). However, if it is the decline in testosterone secretion after the rut that causes the hard antlers to be shed, and the casting process itself and subsequent wound-healing that trigger regeneration and growth of new antlers, how is it possible to explain the situation occurring in those deer species (moose, white-tailed deer, reindeer) in which the antlers are cast in winter but regrowth does not take place until the following spring? Some clue to the process involved here may come from consideration of the antler cycle in roe deer. As in other species, the antlers of older bucks are cast before those of younger animals; in roe this may be in early winter. Regrowth of new antlers is immediate, but at first it is very slow, with growth rate increasing in the spring. It appears that slow antler growth in winter is related to cold conditions and poor feeding. General observation suggests that temperature can have a considerable effect on the growth rate of antlers, with the growth rate increasing during warm weather. Perhaps, in those species which cast in winter, antler regrowth is delayed owing to cold temperature or poor nutrition, with growth initiated or rate of growth increased in spring in relation to increased food intake (Lincoln and Bubenik, 1985).

The cycle of antler growth and casting in male deer is clearly markedly influenced by levels of hormones in the blood, most notably the male hormone testosterone and perhaps also prolactin (Fennessy and Suttie, 1985; Bubenik *et al.*, 1985). Secretion of testosterone itself is influenced by the pituitary secretion, luteinising hormone (LH), and secretion of both LH and prolactin are influenced by day length. The antler cycle in temperate species of deer thus closely parallels the annual cycle of both season and sexuality. In most temperate members of the Cervinae, antlers are grown in the spring (between April/May and August in the northern hemisphere) and are cleaned during August ready for the autumn rut. Hard antlers are retained through the winter and shed in early spring. Antlers are shed earlier among the temperate Odocoilinae of America (moose, white-tailed deer, mule deer, reindeer) when casting occurs during early winter, but patterns of growth are otherwise similar, with development of antlers through spring and early summer, cleaned for an autumn rut. Among temperate species, only the European roe deer reverses the pattern: antlers are cast in November, regrowth begins again at once (but, as we have noted, with initial growth extremely slow) and the antlers are clear of velvet by April or early May, just when the other Eurasian species are shedding their antlers. This reversal, however, makes perfect sense; and it links in just the same way to the sexual and hormonal cycle: for roe deer rut in July and August.

In temperate species, the antler cycle is linked to the reproductive cycle, itself synchronised and linked to seasonal pattern. What of those non-temperate species, which might be considered aseasonal, or at least less strongly synchronised? The factors determining antler growth and casting appear the same, and again seem closely related to levels of male hormones. Moreover, since we noted in Chapter 5 that the breeding of these 'aseasonal' species is not in fact evenly distributed throughout the year but is concentrated into particular periods (even if the range is wider than in species suffering the more marked climatic changes of temperate seasons), it is perhaps not surprising to find that the antler cycle, too, is essentially still seasonal and is related in time in exactly the same way to the different phases of the sexual cycle (see Table 5.1).

If growth of antlers in male deer is controlled by male hormones (at least in the initial and necessary development of pedicles and the retention of the dead bone after growth is complete: pages 137–8), what causes and controls growth of antlers in female deer? While the development of antlers by females is in most species an abnormality (see page 134), in reindeer and caribou antlers are a normal feature in females as well as males. In both sexes, development of the antler pedicles begins very soon after birth. This may mean that pedicle and antler development are not dependent on sex hormones at all as in other species but occur as part of normal somatic development, and may be similar to the way that female sheep and antelope develop horns in some breeds and species but not in others. Such a change from total dependence on sex hormones for antler development could involve a very simple genetic change, perhaps even a single gene (Lincoln and Bubenik, 1985). It is, however, also possible that the induction of antler development in female reindeer is related to an unusually high androgen secretion by the ovary. Attempts to measure the circulating levels of testosterone have shown that in reindeer of both sexes blood levels of this hormone are very low throughout the year. There may, however, be other androgens involved; an alternative source of these may be the adrenal glands. Lincoln and Bubenik conclude that, since antler-pedicle development commences before puberty in female reindeer, it may not depend on induction by sex hormones as in other species. The control of casting and regrowth of the antler in the adult female, however, is probably dependent on a seasonal cycle in the secretion of steroid hormones arising from either the ovaries or adrenal glands.

'Hummels'

One other phenomenon which should be introduced at this point is the occurrence in some species of antlerless males, or 'hummels': males which, although they may possess small pedicles, do not at any stage in their life develop antlers. Such antlerless males appear in all other respects to be perfectly normal; although for many years it was assumed that they must be polled owing to lack of sufficient testosterone (much as, in the occasional aberrant antlered females recorded in roe or white-tailed deer, the occurrence of antlers appears to be related to production of abnormally *high* amounts of this hormone), it would appear that this is in fact not the case. Hummels are rare (perhaps one in 300 of wild Scottish red deer stags according to a survey by Whitehead in 1972), but they appear to be as fertile as antlered males: all red deer hummels studied have testes of normal size, and develop other secondary sexual characters such as a long neck mane, enlargement of the canine teeth and staining of the hair in front of the penis (Lincoln and Fletcher, 1984).

The cause of the polled state is in fact not yet fully understood; because of its rarity there has been little opportunity for intensive research. Once it was established that it was not related in any way to abnormally low levels of testosterone secretion, the obvious alternative explanation for the hummel condition is that it may be a simple genetic mutation, as is hornlessness in cattle and sheep. The only detailed study which has been undertaken, however, reported in a series of papers by Lincoln and Fletcher (1969–1984), suggests that this conclusion, too, is oversimplistic. In breeding trials with antlerless male red deer in Scotland, two hummels were mated with 'normal' hinds to follow antler development in the male progeny. A total of 45 male progeny were produced that survived to two years or more. All these hummel offspring, even those derived from inbred crosses where a hummel stag was mated with his own daughters, grew normal antlers, suggesting no real genetic basis for the hummel state (although, as Lincoln and Fletcher (1984) note, more backcrosses are needed to confirm this beyond doubt).

While Lincoln and Fletcher's hummels lacked antlers, all had pedicles (although these were smaller than those of normal males and appeared never to have properly completed their growth). To test whether the animals in fact had the potential to grow antlers despite the fact that they had never done so naturally, Lincoln and Fletcher inflicted a small wound in removing a small piece of tissue from the top of one of the rudimentary antler pedicles in each animal. The logic of this was to simulate the events which naturally occur at the time of antler-casting in the spring in a 'normal stag'. At this time, it is the wound created on the surface of the living pedicle following the casting of the old antler that leads to the regeneration of the new antler (see above). In view of this, it was predicted that wounding the surface of the antler pedicle of a hummel might result in the growth of antler tissue, and, since hummels appear to have a normal seasonal cycle in testosterone secretion, the new antler tissue should be cleaned, cast and regrown each year once development had been initiated (Lincoln and Fletcher, 1984). This experiment was carried out on three hummels. In two cases surgery was followed by the production of a normal antler on the injured pedicle, and the antler was cast and regrown annually thereafter in the normal way. In each case only one pedicle was operated on and, while on the injured side antler growth was initiated and a normal antler cycle established, no antler growth was observed on the other side of the skull and the pedicle remained rudimentary, making it clear that initiation of antler growth was specific to the pedicle-wounding. In the third animal, however, the attempt to stimulate antler growth was unsuccessful; Lincoln and Fletcher note that this animal was exceptional in that, while, like all other hummels examined, rudimentary pedicles were present, in this particular case they were extraordinarily underdeveloped.

Lincoln and Fletcher conclude that the hummels they studied were not genetically polled, but had the potential to develop normal antlers. They consider that the animals failed to grow antlers spontaneously because their antler pedicles never completed the normal growth. For at least two of the animals this abnormality concealed their true potential, since they were fully able to produce normal antlers, as revealed by the experimental stimulation of antler growth. The failure to produce normal pedicles was not caused by an obvious reproductive abnormality: the animals were of normal body size, had well-developed male secondary sexual characteristics except for the antlers, and were fully fertile although generally subordinate in inter-male rivalry. The animals also showed a normal seasonal antler cycle once antler development had been initiated, indicating that they had a normal season pattern of testosterone secretion from the testes (Lincoln and Fletcher, 1984).

The obvious question is this: why did the antler pedicles fail to grow normally in the hummel stags? The breeding trials provided no evidence of an obvious genetical abnormality, and all male progeny produced by the hummels developed normal antlers. Lincoln and Fletcher (1984) suggest that, at least for red deer hummels in Scotland, the incomplete development of the pedicle results from delayed puberty, caused by malnutrition in early life. To what extent this offers a general explanation for the occurrence of hummel males in other species, too, is of course a matter for conjecture.

EVOLUTION OF ANTLERS

Antlers, while unique to the Cervidae, are not universal within the family. Neither musk deer nor Chinese water deer possess antlers at any stage during their life (and this feature is used by many as an argument that both species should in fact be separated from the Cervidae into their own families: Moschidae and Hydropotidae). Antlers are found in the primitive Muntiacinae (muntjacs and tufted deer), but males of these species retain a canine tusk in the upper jaw like that of water deer or musk deer, and the antlers are simple. As we have noted (pages 16, 135), their pedicles too are somewhat atypical, extending considerably down the face (up to 70 mm in length), which in itself may be considered perhaps a primitive feature; the antlers are usually short (about 60 mm) and are curved in at the tips. Mature males of some muntiacine species may occasionally develop a brow tine (see Figure 2.1).

Simple antlers are retained in some of the Odocoilinae (e.g. the pudu and the various species of brocket deer of South America), but in more advanced species, both within this group and within the Eurasian Cervinae, antlers are more usually branched. In the supposedly more primitive forms of both groups, the antlers are developed on a three-point plan (Geist, 1971; Kurt, 1978) as in roe deer, pampas deer and hog deer. More advanced species within these groups retain the basic three-point plan but show branching from the three main beams (white-tailed deer, rusa, chital and barasingha), while at the highest level of development the antler is based upon a four-point plan, as those of the most advanced members of the genus *Cervus*, the sika and red deer.

Such classification of the deer on antler design is perhaps somewhat artificial, and is not universally accepted (e.g. Gould, 1973; Lowe and Gardiner, 1975). What is generally accepted, however, is that antlers have progressively increased in both size and complexity during evolution. Further, that increase in size is not due simply to a concomitant increase in overall body size. Although it is true that there is a tendency for antler size to increase with body weight (Clutton-Brock, 1982), the relationship is far from linear and antlers become disproportionately larger in the larger, more advanced deer (Gould, 1973). (At the most extreme, in the giant deer or Irish elk — *Megaloceros*, the largest deer that ever lived, but nonetheless an animal which in life stood no more than 6 feet at the shoulder — the antlers had a span of up to 12 feet and may have weighed up to 45 kg: see Figure 2.4.) This allometric increase in antler size in relation to body size during evolution runs parallel with the trend already noted (pages 29, 71–2) to increasing polygyny. As mating systems became progressively more polygynous and male competition for mates thus increased, so antler size is seen to have increased. Nor is the relationship between the two no more than coincidental correlation: among living deer, it is clear that there is a strong positive relationship within polygynous species between reproductive success and antler size (Figure 7.4).

Figure 7.4 The antlers of deer with large breeding groups (solid circles) are relatively larger than those of deer with small breeding groups (open circles), even for equivalent body weight. Source: Kitchener (1987) after Clutton-Brock et al., *(1980)*

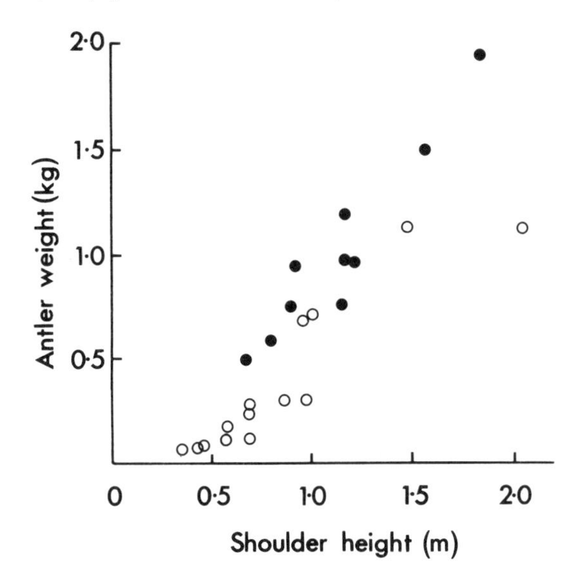

THE FUNCTION OF ANTLERS

The relationship between antler size and reproductive success within species, and, among species, between antler size and degree of polygyny, offers strong hints of the function of antlers. Some species develop only comparatively simple antlers with a single spike, or with at most only limited branching from such a central spike. Such species, however, tend to be the relatively unsocial, often territorial groups: the primitive muntjacs, the solitary pudu or brocket deer, the asocial roe. Their simple antlers may serve primarily as general-purpose weapons, against others of the same species in dispute for territories (page 81) but also perhaps against predators; the development of antlers in reindeer females may also reflect a need for defensive weapons in the face of particularly severe predation and is thus also suggestive of an anti-predator function for antlers in general. In the vast majority of the later deer species, however, antlers have become more complex and their primary function has come to be associated with intrasexual competition.

Males of the more social species compete for females through display or ritualised wrestling matches (page 83). Fights are by no means uncommon, and antlers, being indubitably used in such fights, must be designed to withstand the stresses of fighting. Even when males do engage in combats of this type, however, the *way* in which the antlers are used in fighting has changed (Kitchener, 1987). Thus they are no longer used primarily as stabbing or thrusting weapons, but are interlocked to provide a 'hold' for wrestling matches and pushing contests (Kitchener, 1987). For such a purpose, of course, they have no need to develop the extreme size and complexity that is evident in some species; they need merely to present a sufficient structure to engage the head of other males. So why do species such as red deer, reindeer, wapiti, moose and fallow deer develop such excessively complex antler structure, far beyond what is required for fighting? Clearly, the answer lies in the fact that much male competition is resolved by display before it even develops to the stage where direct fighting may ensue. Indeed, fighting generally occurs only between males which are in other respects fairly evenly matched, and where therefore the contest cannot simply be resolved by display alone. The main features of such display, outlined on page 83, are parallel walking, a strutting walk side by side with antlers prominently displayed and head erect, or face-to-face confrontations with antlers lowered and swinging slowly from side to side. Antler size is related to body weight, and body weight is in turn related to dominance; thus antler size may be expected, at least to some degree, to reflect dominance and it is clear that the antlers play an important role in such display (Bubenik, 1968; Geist, 1971; Topinski, 1974; but see Clutton-Brock *et al.*, 1982). In such case, natural selection might favour an increase in antler size beyond that minimum required for actual fighting.

Indeed, there is much controversy as to whether, as antlers have developed greater and greater complexity and greater size, their function has shifted completely from that of an actual weapon to become organs of pure display, so that, at the extreme, they retain no function in direct conflict. (It is argued by many, for example (after Gould, 1973), that the antlers of the ultimate extreme, the Irish elk, could have been of no possible use in fighting.) The truth probably lies, as usual, somewhere between the two: for there is little value in display unless its connection with actual fighting ability is occasionally reinforced with a little demonstration and, as we have noted, display alone cannot resolve symmetric contests between apparently equally matched individuals.

The actual structure of even the most complex antlers also bears out this idea that they must retain some fighting function. In almost all species of deer with branching antlers, the first branch from the main beam, the brow tine, projects forward over the eye (see Figure 7.7). When two competing males interlock antlers they lower their heads at an angle, thrusting the main branches of the antler forward to engage with those of the rival. In this position the brow tines are never long enough to reach the opponent, and are of no real value as weapons. They do, however, cover the eye of the owner and may be designed, as the foil or guard on a rapier, to protect the eye from injury (Figure 7.5). The shape of the rest of

Figure 7.5 The fighting positions of living deer: (a) reindeer; (b) red deer; (c) swamp deer; and (d) chital. In all cases the antlers are engaged towards the base so that the terminal tines point directly at the opponent; the basal, or brow, tines protect the eye. Source: Kitchener (1987); reproduced by permission of Modern Geology

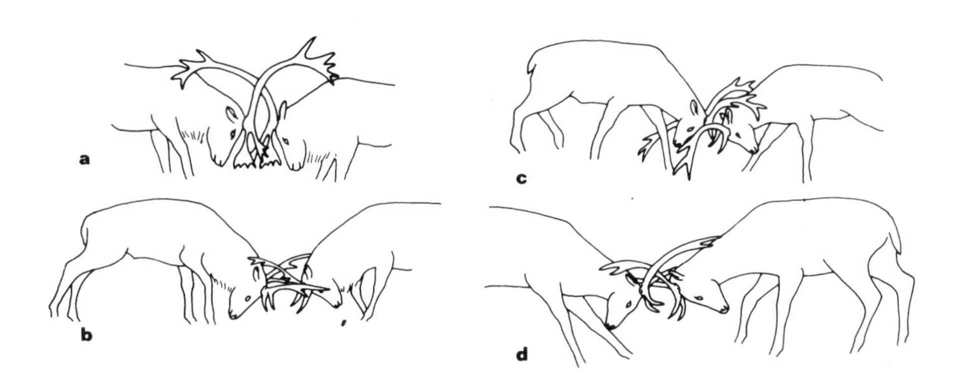

the antler, too, seems designed specifically to facilitate interlocking (Kitchener, 1987).

Finally, not only the shape but even the mechanical structure of antlers is suggestive of the retention of some function in actual combat. The cross-sectional shape of the main beam changes at different points along its length, at each point presenting the optimal structure to withstand the pushing and shearing stresses which would be experienced in an actual fight, and reflecting these more precisely that would be expected were the structure merely to support a given antler shape and weight. In those species in which the upper part of the beam is flattened, the breadth and orientation of the palm also seem adapted to accommodate the forces to be expected in combat. Nature is in general economical, conservative. The growth of antlers is at best expensive, and it would seem unlikely that any animal would devote more of its limited resources to development of an antler than absolutely necessary. Indeed, natural selection would act quickly to eliminate such wastrels, with the result that antlers would be sufficient — and no more — for their intended function. Yet the strength of palm, tines and main beam in all species, including the Irish elk, is far greater than that needed purely to support the actual weight of the antler itself as a pure display structure. In fact, the mechanical strength of the antlers of the Irish elk is shown by Kitchener (1987) to have been more than 67 times that which would be required for this purpose. That biomechanical strength is achieved only by the deposition of extra bone around the antler beam, many times more than really necessary. This cannot make sense and is clear evidence that the antler must have more than a purely passive role to play. Calculating the actual stresses which would be imposed on antlers of various species in actual combat, Kitchener has shown, however, that in all cases antler strength is only just sufficient to withstand these stresses: the safety margin is now a realistic 1.7 rather than a massive 67, far more in keeping with the conservatism of natural selection.

All such evidence points to the retention of some fighting function even among those species whose antlers have developed beyond the minimum structure required for fighting alone. Such considerations may also offer an explanation of why antlers are — apparently so wastefully — shed and regrown each year. If they have come to assume such importance in display and not just in fighting, and yet (because they still are also used in actual combat) they are subject to possible breakage and injury, it may be important to replace them each year: not to increase their size (which could be achieved by simple accretion to a permanent structure, as is the case in the horns of bovids), but to repair any damage caused in the previous year which may reduce their *apparent* size and thus the dominance they may secure their owner in future display.

There is, however, another possible explanation for the deciduous nature of antlers — a passive consequence of their structure and physiology rather than an active, evolutionary device relating to their function as organs of display and combat. Antlers could simply be a parallel adaptation among cervids to the horns of bovids, different not for any special purpose but merely through accident of evolutionary chance. It is quite possible that both horns and antlers were developed by, respectively, bovids and cervids quite independently of each other, but in response to analagous selection pressures; developed for a similar purpose, but by evolutionary happenstance from a different origin. In this case, antlers are not different from horns because they need to be, but because of a rather different evolutionary origin. Even so, as a consequence of that origin, and as a function of the curious physiology of their growth, it is nonetheless essential for antlers to be cast each year.

Development, progressive calcification and thus ultimate death of the antler when fully developed appear to be under the control of the hormone testosterone (page 139). Besides causing the death of the antler, testosterone is also involved in the retention of this dead tissue through the breeding season; when levels of testosterone fall at the end of the season, the antler is cast. In addition to all this, however, a further process is occurring. When antler growth is complete, progressive calcification causes death of the antler as we have noted, but there is also a gradual die-back of the tip of the pedicle at the junction between pedicle and hardened antler, a die-back that continues as long as the antler remains in place. When the blood level of testosterone declines in the spring, the old dead antler is cast and a new one regenerates from the wound created on the surface of the pedicle. If for any reason the dead antler is not cast (as, for example, when this is prevented artificially by maintaining a continuous high level of testosterone in the blood stream with hormone implants), the die-back at the junction between antler and pedicle continues down the pedicle and into the skull — and can be fatal (Lincoln *et al.*, 1972; Fletcher, 1975). This progressive die-back of the pedicle after antler death thus provides another compelling reason why dead antlers must be cast each year and not retained, even if there is no more 'active' purpose.

THE SHAPE OF ANTLERS

While the fine detail of the design of antlers may be influenced by mechanical considerations of supporting the weight of a display structure and withstanding the stresses experienced during fighting, there is clearly much variation in basic shape among the different species of deer. Thus, as we have already noted, antlers may be straight or curved: that is the main supporting beam may be erect and straight, as for example in roe deer; or curved, either backwards and upwards as in sambar or reindeer, or sideways and up as in, for example, the moose. Further, they may be simple or complex; while juvenile males in their first 'head' generally support only single spikes and more complex heads are developed only in later years, in some species even mature bucks retain a simple antler, although it will increase in weight and length in successive heads. In those species which do boast more complex structures, as we have seen, these may be based on three or four main branches. Also, although normally the main beam assumes dominance, so to speak, running the full length of the antler, with all other branches departing from it at different points along its length (e.g. Figures 7.2 or 7.7), this is not inviolable. In mule deer, the antlers branch dichotomously — that is at each division the two branches diverge at equal angles from the direction of the previous 'stem'; and in Eld's deer (the brow-antlered deer) the first branch or brow tine assumes dominance (Figure 7.6). Although antlers may be seen to fit a one-, three- or

Figure 7.6 The elongated brow tines of Cervus eldi *give it its common name of brow-antlered deer*

four-point plan, these main branches may themselves support additional 'tines' (terminal points developing from the main branches). The absolute number of tines varies tremendously, from the simple spikes of muntjac or pudu, through the three-pointed antlers of roe deer, to the antlers of reindeer which may have in total more than 44 points. Finally, in some species the upper part of the main antler beam is flattened and broadened to form a 'palm', from which terminal points branch as 'spellers'. Palmation of the antlers is seen in moose and in fallow deer and was of course characteristic of the great Irish elk, and we may wonder what prompted its development in these apparently unrelated species and why, if it is so to speak an option, only these three species did develop it.

In fact, despite their apparent variation, antlers of most species are based on the same general plan, a fact implicitly recognised by virtue of the fact that the special 'antler jargon' developed by sportsmen and hunters to describe antler shape and 'quality' is readily applied to any species. Thus a main *beam* is identified of a given width and cross-section; the first, forward-pointing branch (and usually the longest) is referred to as the *brow tine*; that above it the *bey* or *bez*; the next the *trey* or *trez*, and so on (Figure 7.7). If the main beam ends in a crown or cluster of tines all

Figure 7.7 The correct terminology for the different parts of the antler

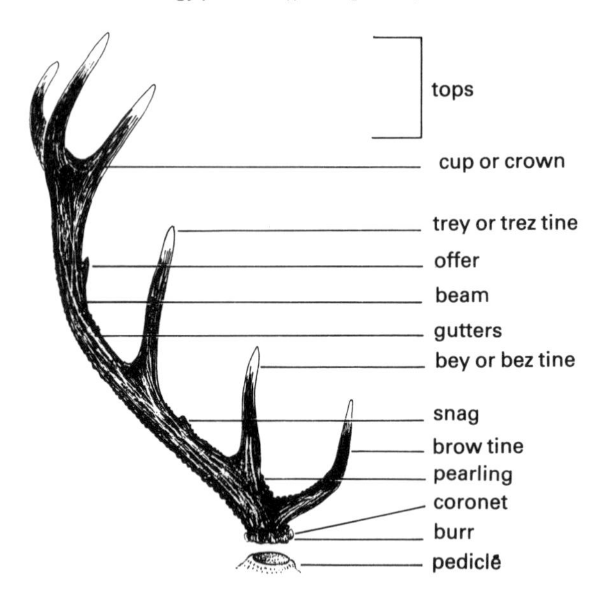

arising simultaneously from one point, or spreading from a broad palm as in moose or fallow deer, these are regarded as 'spellers', and it is clear that in the most general terms antlers of all species do accord to the same basic pattern. (Interestingly enough, while the entire appendage is now referred to as an antler, and thus modern terminology distinguishes it most strictly from the non-deciduous *horns* of other ungulates, this is a relatively modern idiom. In the 'Sportsman's Dictionary' of 1735, some 250 years ago, the 'attire' of a deer was described quite firmly as a pair of horns. Each horn was formed of a beam, and the lower two 'starts' (or branches of the beam) were called antlers: brow antler and bes (sic) antler respectively. Thus, while it was acceptable to speak of a deer's antlers, 'any self respecting hart', as George Darwall (1985) expresses it, 'will have more than two of them'. In the same terminology, incidentally, all others 'which grow afterwards until you come to the crown, palm or croche are called royals and sur-royals; the little buds or broches about the top are called croches': Darwall, 1985.)

While the 'ground plan' of antler design may thus be essentially similar in all cases, there is nonetheless considerable variation expressed upon this theme, as our survey above makes clear. In addition, not only is there wide variation in antler form among species, there is also much variation among individuals, so that it may be claimed with some validity that no two pairs of antlers are exactly alike — and it is certainly possible to recognise an individual male from the shape of his antlers. With all this variation, what does in fact determine antler shape? What sets the initial ground rules to which all must adhere, and what results in the variations around that? Darwall and Clark (1986) have tried to identify the general rules of antler design. They note that, for example, tines tend to curve upwards rather than downwards; in most species, while the first branch or brow tine projects forwards, the main beam sweeps back, with other branches projecting forwards or outwards from that beam. Growth is predominantly within a smoothly curved surface in space: that is to say tines do not protrude randomly, and never (except where the antler is damaged or malformed in some way) project inwards. Thus, the internal surface, down the inside of one antler, across the top of the head and up the inner surface of the opposing antler, is always smooth.

In a fascinating analysis Darwall and Clark conclude that, despite the great diversity of antler form we observe, there are a number of these common principles of design, and there is a relatively simple explanation for the basic shape that antlers assume. Their model of antler growth envisages particular growth points, which are distributed over the surface of the pedicle after casting. It is possible that further growth points, not originally on the pedicle, may arise during the course of antler development. At least one growth point is required for each tine in the final antler. These growth points remain at the surface of the developing antler (i.e. in the velvet) throughout growth. New cells are formed by mitosis centred on the growth points, and the cells spread out over the surface of the existing antler under the velvet. Ossification of this tissue produces a new surface over which further cells can be spread in turn. This mechanism gives rise to the cap of dividing cells which can be observed. Each growth point generates such cells independently, except that the rate of mitosis at any one growth point may be affected by the presence of others nearby.

Darwall and Clark proceed to develop a computer model to 'grow' antlers according to these simple rules. In order to define the growth process completely, to give a deterministic model, some additional data or assumptions need to be made. The exact details of these assumptions, however, are not fundamental to the conclusions reached. The data required concern: the initial shape and size of the pedicle; the distribution of growth points over the pedicle; the manner in which cells spread out over the surface of the antler in velvet; the rate at which mitosis occurs at different stages of growth; the manner in which cell division is affected by

Figure 7.8 Computer simulations of antler growth. Where several growing points are considered on a flat pedicle, differences in the number and spacing of those focal points result in construction of 'antlers' similar to those of (a) a roebuck (Capreolus *sp.) and (b) a fallow deer* (Dama *sp.). The dotted lines represent successive incremental stages in growth. Source: Darwell and Clark (1986)*

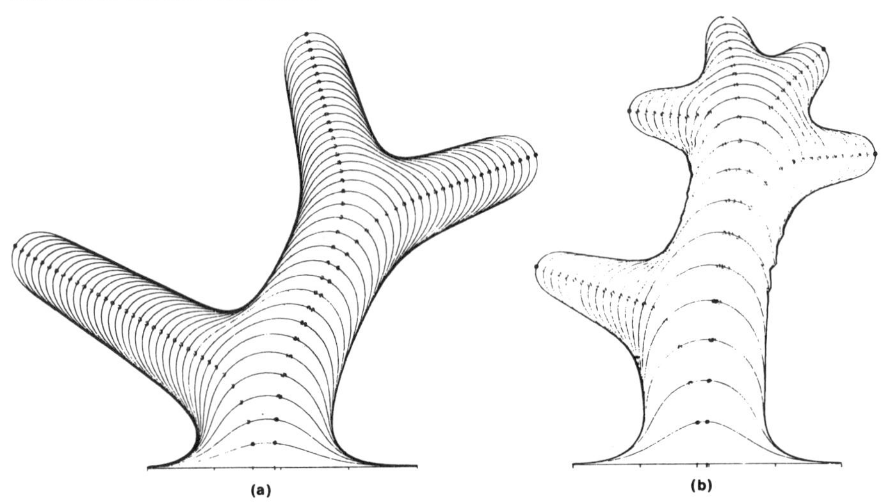

the proximity of neighbouring growth points; and the mechanism (if any) for generation of novel growth points during development (Darwall and Clark, 1986). (Many of these features could be at least partly under genetic control and could provide a basis for species-specific antler shapes.)

From particular but different distributions of growth points within the pedicle, Darwall and Clark's computer model produces crude approximations to antlers reminiscent of roe buck or fallow deer (Figure 7.8). More generally, all the 'antlers' they produce obey the various general principles of shape and form outlined above. The observation, for example (Thompson, 1942), that antlers tend to lie in a smooth surface in space has an interesting consequence. If it were accurately true, then the growth points at initiation of antler growth would all lie in this surface. By converse, if the growth points do lie in a smooth curve at the pedicle, then the resultant antler, not just in its final form but at all stages of growth, must produce a smooth surface. If growth points in the antler are not evenly distributed along this smooth curve, but are uneven in spacing, the tines will form at different points and different spacing from the main beam. Likewise, if growth is asymmetrical about the focal point, the effect will be to make the tine curve. Gravity has been suspected of affecting antler morphology (Holmes, 1974), and could cause such asymmetry: if more cells were concentrated on the 'downwards' side of a sloping antler surface, the computer simulations suggest that the tine would curve upwards in consequence. Other possible agents for asymmetry might be nutrient or hormone supply, or effects due to interactions between close growth points which have yet to diverge. Finally, Darwall and Clark's simulations offer an explanation for the normal backward sweep of the main beam of branched antlers. Quite simply, if an antler develops a forward-pointing brow tine, the rest of the antler is automatically thrown backwards.

Darwall and Clark's analyses suggest simple explanations for many of the general characteristics of antler design; indeed, show those same features as inevitable. In addition, their model offers a simple mechanism explaining or accounting for the variations observed among species in antler shape. The number of tines ultimately developed is a simple consequence of the number of initial growth points on the pedicle; likewise, the distance between branches reflects the linear distance between adjacent growth points in the antler's origin.

INDIVIDUAL VARIATION IN ANTLERS

We have noted that even within a species there may be marked variation in antler form and size. Antlers generally increase in size and complexity with age (page 136), but even in two animals of the same age antlers may not be equally well developed. Since antler trophies are the main preoccupation of those who stalk deer for sport, there has long been great interest in what factors 'make' a good head, and bitter controversy as to what is genetics and what environment. Within the general design characteristic of the species as a whole, precise antler shape is indubitably hereditary to a degree. Offspring of particular males inherit particular tendencies: for a wide spread, a particular arrangement of tines, etc. Thus, among red deer devotees, stags originating from the captive, park herd at Warnham, in Sussex, England, are instantly recognisable by the wide, almost horizontal, spread of the main beams and the upward curve of the terminal tines — almost like bicycle handlebars. Roebuck from different populations in southern England can easily be distinguished by the distance between the pedicles across the top of the skull and by the deviation from the vertical of the antlers themselves (whether they grow straight up, parallel to each other, or diverge to a degree). The size and complexity of the antlers developed, however, are equally dependent on nutrition. We have commented on the incredible demands, particularly in terms of minerals, placed upon a male deer in the brief period of antler regrowth. If minerals are limited in supply, or if the level of nutrition in general is sufficiently lowered that the animal cannot so easily afford even the basic energy costs of antler growth, smaller antlers are produced. The weight of antler produced by four-year-old red deer stags maintained in captivity, at different levels of food intake (Fennessy and Suttie, 1985), is clear evidence of this effect of nutrition (Table 7.1). It is clear from a variety of such studies that both genetic and environmental factors will act to determine the eventual size of antlers achieved.

Table 7.1 Antler weights of captive red deer stags maintained on different levels of nutrition (data from Fennessy and Suttie, 1985)

Feed intake	Antler weight
Unrestricted food intake	2.20 kg
Intake restricted to 80% of that consumed by unrestricted group	1.87 kg

One further implication of this is that it is not possible, as many deer-managers might wish, to predict the 'quality' of antlers a male may have as a mature animal from the relative size (in relation to its peers) of its first or second set of antlers. It has long been assumed that the type of antlers grown by an animal early in its life will indeed be an indication of its future potential, and many deer-managers have attempted to improve the antler quality of a herd by selective culling of 'poor' two- or three-year-olds. It might be argued that, even if antler growth is influenced by environmental factors as well as by heredity, all animals in a given population will be experiencing similar environmental constraints, and thus, where an animal appears to be poor in relation to its peers, this must be due to the genetic component of the relationship. This is, however, to overlook social factors, to overlook the fact that within a deer population there is always a distinct dominance hierarchy. Animals of a given age group will not necessarily have equal access to food or to food of equal quality; level of nutrition may well be related to dominance rank, and if dominance positions change from year to year then the plane of nutrition experienced by an individual in relation to that of its peers may also change. Dominance hierarchies among mature males are fairly well fixed and subject only to little variation; among younger males, however, dominance positions are in constant flux. Whether this is sufficient explanation or not, there is no doubt that recent studies, in which individually-known males have been followed from birth to perhaps eight or nine years of age, show clearly that there is no relationship between the antler size of an animal in its first or second head and the size of the antlers it will produce at maturity.

8 Relations with Man

Man has always assumed a particular interest in wild deer, hunting them for meat, hides or antlers or other commodities, hunting them for sport, persecuting them as pests or as competitors with his domestic livestock. In fact, as we suggested in Chapter 6, there are probably now relatively few populations of deer in the world that are not influenced, and strongly influenced, by Man. Nor is the impact of human activity restricted to the 'predatory' role implicit in that context — in assuming responsibility for control of population size. Human interference has had a dramatic effect on the worldwide distribution of many deer species in the first place: through deliberate introduction of species into new locations, or threatening others with extinction or near-extinction through habitat erosion or over-exploitation.

HUMAN INFLUENCE ON GEOGRAPHICAL DISTRIBUTION OF DEER

Translocation of deer and their establishment in new areas has continued through history into the present day. Man's influence on distribution patterns is indeed so profound that it is for some species quite difficult to tease out what would be their 'natural' biogeographic distribution. As a family the deer are essentially animals of the northern hemisphere, occurring naturally south of the equator only in South and Central America. Yet, in the present day, reindeer are to be found on many antarctic islands where, in the 1920s and 1930s, they were established to provide food for visiting whaling vessels; red deer, wapiti, moose, sambar, rusa, sika, fallow and white-tailed deer have all at various times been introduced into Australasia, with introductions of all species into Australia and New Zealand continuing even as recently as 1910–20. Even within the northern hemisphere, translocations have been commonplace; indeed, some species owe so much of their present-day distribution to human introduction that their geographical occurrence today most elegantly reflects a global history of trading routes and trading partners. Fallow deer for example are believed originally to have occurred in the Mediterranean zone within Europe and from Turkey to

Iran. Widespread throughout the region some 100,000 years ago, they probably became extinct during the last glaciation, with the exception of a few refuge populations in southern Europe. Yet, from these relict populations, fallow have now become established in some 38 countries, between 61°N and 46°S (Chapman and Putman, 1988). The species is now established in the wild throughout Eurasia, in Africa, Australasia and in both North and South America. Lowe and Gardiner (1975) similarly suggest that many of the forms of sika deer of the Asiatic mainland are of introduced and/or hybrid origin, with their distribution and genetic 'provenance' the direct result of Man's ancient trade (see also page 33).

Even today translocations are still going on, although in general such blatant interference is effectively restricted to the reintroduction of species into areas of the world in which they have relatively recently become extinct (e.g. the current reintroduction of Père David's deer into China).

What has prompted such widespread introductions? Clearly, modern *re*introduction programmes are directed towards conservation, but the reasons behind introductions in the past are many and varied. Many species have been brought home by explorers or traders to foreign lands as curiosities or novelties, to become established in private menageries or collections of emperors and nobility. Escapes from such collections would have been frequent, and where conditions proved suitable feral populations rapidly became established. It was in this way that Chinese muntjac and Chinese water deer originally became established in Britain, escaping from the private collection of the Duke of Bedford at Woburn Abbey. This is, however, neither a modern nor an entirely British speciality: the Roman nobility, the Phoenicians, and many of the Asian potentates maintained collections of exotic species, from which feral populations may have sprung. Further, whereas in the West, in Europe, such collections were essentially maintained for curiosity, in the East they were much more functional in objective. The velvet covering of growing antlers is much prized for medicine throughout Asia (see pages 157–160), and many of the captive herds of deer held in this region were established to provide a controllable source of this prized commodity.

While escapes from such collections, both in Europe and in the East, must indubitably have been commonplace, the establishment of feral populations from collections such as these was not always so accidental. In many cases animals may have been deliberately released, or even introduced into the wild in the first instance. Thus, while muntjac may first have become established in the UK as escapes from the Duke of Bedford's estates at Woburn, there is no question but that there were subsequently a number of quite deliberate releases of additional animals. Sika deer, too, while perhaps initially established in Britain as escapes, were also deliberately released in a number of areas (Ratcliffe, 1987). Moreover, introductions of Eurasian species into Australia and New Zealand were unashamedly directed specifically at the establishment of populations in the wild; the animals were released immediately on arrival or very soon afterwards.

Such deliberate introductions into the wild may also have been purely for curiosity value, to increase the diversity of deer species that might be encountered in any area; in Australia and New Zealand, perhaps, to establish familiar animals in an unfamiliar world almost as an act of sentimentality, of nostalgia, to re-create a 'Little England' for the British settlers. More generally, however, deer populations have been introduced as game for hunting: as a source of meat (reindeer in the Arctic and Antarctic islands, or the rusa deer introduced into New Guinea) or for sport (the introduction of Eurasian species into Australia and New Zealand). Sporting interests also underlay the European enthusiasm for introducing new species into areas which already had a native deer fauna: the introduction of exotics provided a new challenge, or offered increased diversity of quarry to the sportsman.

It is worth noting in passing that such introductions are not always crowned with success. Although many deer species have shown themselves highly adaptable and establish feral populations in new areas only too easily — to the extent that they regularly develop later into a major pest problem — there are exceptions. Both Indian and Chinese muntjac have frequently been kept in captivity in the UK: both have certainly been released into the wild, either deliberately or accidentally. Perhaps because of climatic considerations, however, only the Chinese or Reeves's muntjac has actually become established as a feral species (Corbet and Southern, 1981). Similarly, although ten species of deer were actually introduced into New Zealand in the late 19th century, only eight species became established successfully in the wild (page 20), and only red deer and fallow are common. Across the water in Australia, releases of deer by the various 'Acclimatisation Societies' devoted to the introduction of wild animals and birds included swamp deer, Japanese and Formosan sika no longer recorded in the wild, as well as the six species now established as ferals (red, fallow, sambar, rusa, hog deer and chital).

THE EXPLOITATION OF DEER BY MAN

With the exception perhaps of curiosity value or sentimentality, these varied reasons underlying the introduction of deer across the world are essentially the same as those which have in general resulted in the *exploitation* of deer, whether of exotic or of native species: to obtain meat, hides, velvet or other products (hard antler for carving or for medicine; musk, etc.), or to provide sport. Deer have been exploited for centuries for meat and hides, for antlers to use as picks or tool-handles, sinews for rope, by hunting peoples in any part of the globe where they occurred. As life styles changed with the coming of agriculture, the reliance on deer as a primary resource lessened in most areas; but the hunting continued, whether for sport or for reasons of control where the wild deer now threatened Man's livelihood by raiding his crops.

In some areas, actual reliance upon the local deer population continued: notably up in the high Arctic, where agriculture at best remained precarious, at worst impossible. So the peoples of the far north of America continued to hunt caribou during their seasonal migrations up into the tundra, taking as many as the short season would permit before the animals moved on again in their ceaseless drifting through the continent; these were to provide meat, which could be smoked or dried, hides and rope to tide them through the rest of the year. In northern Europe the relationship with reindeer became still more intimate, still more fundamental. The Lapps relied upon the reindeer for everything: meat, skins for clothing and for building their flimsy shelters, antlers to be shaped into tools, and tallow for light. During the late summer and autumn, reindeer milk was available; although the quantities are small (about half a pint a day), the milk is rich in protein and fat, with a butter fat content on average of some 20.80 per cent (with a range of 18.75–22.95 per cent). The milk taken could be drunk fresh or turned into butter, cheese or yoghurt. So dependent did the Lapps become upon their reindeer herds that they followed them through their migrations (more local in terms of distance than those undertaken by the Barren Ground caribou of North America). They came to dominate the herds; so that they might exercise control, they caught and tamed young calves which were later returned to the wild herds as lead animals whose comings and goings could be ordered by the Lapps themselves. Increasingly the great reindeer herds became more and more subject to Man's control, finally becoming effectively semi-domestic as they remain today.

Such total dependence on the deer herds was, however, exceptional. Throughout Europe and Asia, the economy of life became more and more supported by agriculture. Deer were viewed as vermin, as pests — or as game for sport-hunting. The pursuit of deer became almost universally the privileged sport of the nobility. Romans, Egyptians, the people of the Indian subcontinent, China and elsewhere throughout Asia, as well as Europe, all became obsessed with the hunt. Great traditions arose as to how each species should be pursued, strict rules governing the conduct of the hunt itself. The whole became surrounded with ceremony, pageantry and tradition. Strict seasons were laid down when each species might be taken, avoiding the summer months of calving, and the rut; certain species were set aside for monarchs, others for lesser nobility. The whole reached perhaps its peak of extravagance in mediaeval Europe. Of the European deer, red deer, the Royal beasts of the Forest, were reserved for the Crown (along with the wild boar, the hare and, at least until 1338, the roe). The Royal privilege was not restricted to the monarch's own properties but extended throughout the land. Wherever an area was deemed to come under Forest Law, as distinct from Common Law,* then, whoever might actually own the land (whether the King himself or some other), Royal game within that area was the King's preserve and no one but the King (not even the landowner) could hunt the Royal beasts without dire consequence. Under the Norman kings of England, death was the penalty for killing a red stag, and it was not until the reign of Henry III that the laws were relaxed sufficiently to decree that no one should lose life or limb for killing Royal game. Such restriction was perhaps recognised as somewhat over-prohibitive: to sugar the pill a little, Royal permission was granted to some noblemen to create 'Chases' on their private estates. These were in effect the hunting preserves of the lesser nobility, but even here, as we have seen, they could not hunt the Royal game without the King's authority. Instead they were restricted to the fallow deer, the fox, and latterly the roe: the 'beasts of the Chase'.

---

* This reservation of certain areas to come under quite separate Forest Law, as distinct from Common Law, is in fact the original meaning of the term Forest: as signifying areas of land to be considered the hunting preserve of the Crown. The present-day equation of Forests with trees is a relatively recent, and purely secondary, association.

In recognition of such complex demarcation the different species of deer were given distinctive names: bucks and does for males and females of fallow and roe, stags ('harts') and hinds for the Royal red deer (page 29). In addition, as described in Chapter 2, each age-class of male was dignified by its own distinctive name. The whole hunt, too, was graced by its own language, with the spoor on the ground referred to most carefully as 'slot', and with droppings honoured as 'fewmets'. Hunting was in truth more than mere sport; it was raised to the level of ritual.

Of the old hunting tradition, little now remains beyond the language. Today, only in very few places are deer still hunted on horseback, with hounds. The killing of deer for sport, however, most certainly has not disappeared; it has merely changed. Hunting is as popular today as it was in mediaeval times, but today's hunter stalks his quarry on foot, armed with a high-velocity rifle. Some of the old tradition lingers: today's stalkers still preserve the old language, still maintain the same distinctions, and still prize the deer stalked in relation to its age and antlers. In the mediaeval age, stags or bucks were graded on the size of antlers and the number of branches. Each branch or tine had its own special name (see Figure 7.7); and there were special terms for antlers with more than a certain number of points. (Thus, in Scotland, a stag with ten points to his antlers becomes a 'royal' stag — originally, 'fit to be taken by royalty'— one with 12 points an 'imperial', and so on.) This much is preserved today, and the 'trophy' value of the head is of paramount importance to the stalker. Indeed, complex formulae, based not only on the number of tines but on the spread and breadth of beam, are now used, and trophies are awarded bronze, silver or gold medallions by the various sportsmens' associations.

If some disapprove of hunting for sport, it is only fair to point out that the interest generated by hunters often acts to focus attention upon the deer and to ensure legislation and management to secure their well-being: exploitation can often be a strong incentive for conservation. In addition, in many countries, the letting of stalking rights or the licensing of hunting provides an important source of revenue.

While in the West, with a diminishing need to rely on deer for meat, hunting shifted in its emphasis towards killing for sport, or to control deer as pests, rather than in a genuine *exploitation* of the deer as a resource, in the East meat was never the only major product supplied by the deer. Deer and deer products have for more than two millennia played a central role in China and other parts of Asia as a source of medicine, and still do so. Despite development of a more agricultural economy and a lesser reliance on the deer as a food animal, therefore, hunting of deer in the East has continued, up to the present day, to be in exploitation, as a source of these essential medicines. Until recently, the uninformed West imagined that the Eastern appetite for such deer products was based on a belief that they had aphrodisiac properties, but the truth is far from this, and deer products are instead widely used in medicine to treat a variety of ills.

The earliest written record of the medicinal value of deer products was found on a silk scroll recently recovered from a Ham Tomb in Hunan Province, China. The scroll chronicles several significant medical treatises, including a section on prescriptions for 52 diseases; in three of these prescriptions, deer antlers, venison and a glue prepared from deer

Figure 8.1 A page from Chêng Ho Pênts'ao, *the official* pênts'ao *of the Sung dynasty (published AD 1116), showing a spotted deer or chital. The text printed in negative is an excerpt from an earlier* pents'ao: Shên Hung Pênts'ao Ching, *from around AD 200. Source: Kong and But (1985)*

antlers were included. More systematic recording of the medicinal virtues of deer parts was compiled in a series of Chinese herbals, collectively known as *pênts'ao* and dating from AD 200 (Figure 8.1). Altogether, 25 parts of deer are registered in *pênts'ao*: as velvet, antler, antler glue, bone, bone marrow, spinal cord, penis and testes, venison, headmeat, head glue, sinew, blood, tooth, shank, skin, fat, brain, semen, gall-bladder, thyroid gland, meconium, foetus, undigested milk and bone lower limb (review by Kong and But, 1985). To this list have been added three more parts: tail, stomach and gastrolith (Hsu *et al.*, 1979).

The claimed cures range from treatment of general debility or 'malaise' to the more specific relief of symptoms related to consumptive diseases (vertigo, coughing, palpitation, insomnia, and tinnitus), spermatorrhea, 'wet dreams', impotence, lumbago, gonalgia, frequent urination, metrorrhagia, persistent bloody discharges, leukorrhea, amenorrhea, abnormal menstruation, retained afterbirth, infertility, habitual miscarriage, carbuncles, dermatitis, traumatic injury, goitre, hemoptysis, epistaxis, apoplexy, epilepsy, diabetes, fever, and neurasthenia.

Dealing with as many deer parts as these and their reputed functions, the *Pharmacopoeia of the People's Republic of China* (1977) adopted a more cautious stance, registering only four deer parts (namely, the antler, antler glue, residue of antler glue, and velvet) and recognising their functions only in the treatment of lumbago, gonalgia, mastitis, ecchymosis, carbuncles, turberculosis in bones and joints, impotence,

Figure 8.2 A variety of over-the-counter preparations available in Asia, based on deer products. Source: Kong and But (1985)

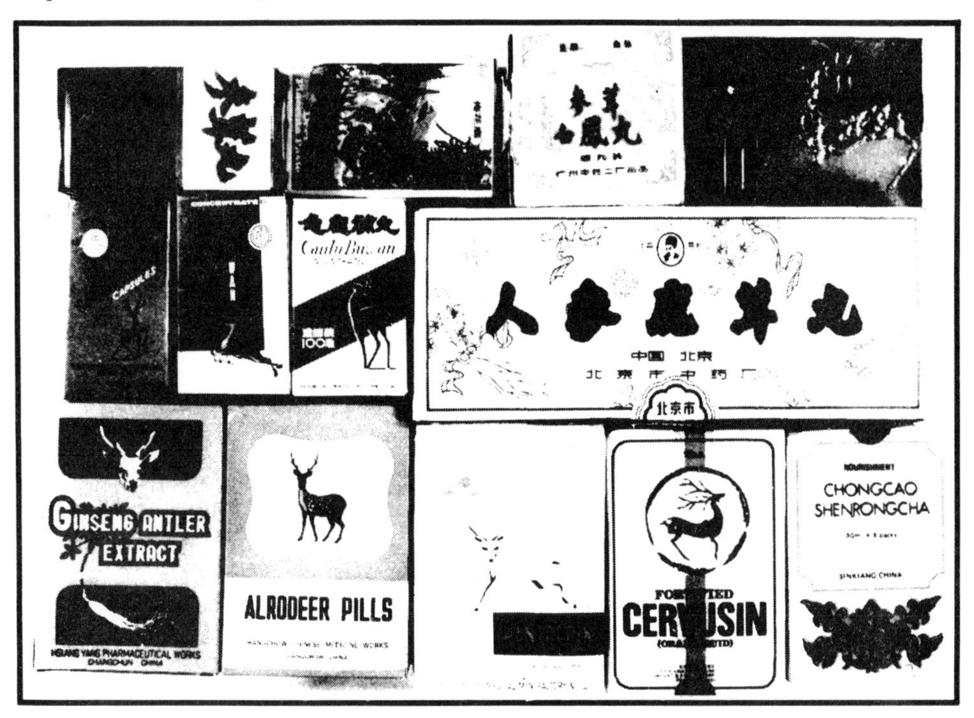

spermatorrhea, metrorrhagia, frequent urination, 'wet dreams', vertigo, and anaemia. The general public, however, holds a firmer belief in the value of the various deer parts. A compilation of the prescriptions of over-the-counter drugs used in China showed that deer parts were incorporated in 76 kinds of drugs, which can be classified into five groups covering 48 tonics, 23 drugs for the treatment of gynecopathy, three for rheumatism, and one each for gastro-intestinal problems and cardiovascular problems (Kong and But, 1985) (Figure 8.2).

The cynical western mind may look askance at such ideas, and dismiss them as pure folklore. It is only fair to note, however, that rigorous scientific testing has been undertaken recently of the efficacy of some of these various preparations, and, although in many cases it is clear that claims made for their curative powers are not fulfilled, in other cases there is evidence that the drugs may indeed have some real pharmacological properties (even though it may not be clear from biochemical analysis why they should work): see Kong and But (1985).

Nor is the trade in deer products a minority, specialist trade. Some measure of the quantities of materials used in China alone can be gleaned from figures for imports and exports through Hong Kong, which because of its strategic position acts as a major clearing house for deer products in the East. In 1981, according to Kong and But, nearly 107 tonnes of deer parts (excluding venison), valued at about 54 million Hong Kong dollars (or £4.5 m), were imported.

One other deer product worthy of special mention here is musk. For centuries, the musk deer (*Moschus moschiferus,* and other *Moschus* species) has been exploited for its musk, a granular secretion of the preputial gland in males. This gland is differentiated in the musk deer into two regions: an outer, glandular region, concerned with the secretion of the musk, and a central sac in which the musk matures and is stored. Males start to secrete musk from the age of 12–18 months; peak production is between May and July and is marked by a visible swelling of the sac and scrotum and by a loss in appetite (Green, 1985). This condition lasts for about two weeks, during which time a yellow milky fluid drains into the neck of the musk sac; here over the next four weeks or so it matures into a powerfully scented, red-brown, granular substance. In mature males the sac contains about 30 g of musk, occasionally up to 45 g, but in young males the 'pods' as they are called may contain only 5 g or less (Flerov, 1952).

Musk is one of the oldest of raw materials used in perfumery, and is highly esteemed on account of its fixative and scent properties (Green, 1985). Its use dates back at least to the 8th century, but in modern times very little musk is used in perfumery owing to its cost and to the difficulty of obtaining the genuine product. Most natural musk is used today in Eastern medicine, as other deer products we have described, although use of musk is something of a Japanese speciality: Japan accounts for some 85 per cent of the entire international trade in musk. It is used as a sedative to treat asthma, epilepsy, and other nervous disorders, and as a stimulant to cure bronchitis, pneumonia, typhoid, typhus and other ailments (Mukerji, 1953; Chopra *et al.,* 1958). Nor is its efficacy in treatment of such ills pure folklore: various workers have shown that musk is effective as a general stimulant of the heart and central nervous system (Mukhopadhyay *et al.,* 1973) and as an anti-inflammatory agent (Mishra *et al.,* 1962; Seth *et al.,* 1973; Taneja *et al.,* 1973).

Musk is more valuable than gold and is reputed to be the most expensive animal product in the world, fetching up to 45,000 US dollars a kilo on the international market (Green, 1985). At the turn of this century, when the musk trade was at its peak, annual exports from China and the Indian subcontinent totalled around 1,400 kg, although today the quantity has dropped to some 300 kg a year (Green, 1985). (It is important to note that this decline is not due to a decrease in demand for musk, but rather reflects the rapidly diminishing populations of musk deer remaining in the wild, decimated by the musk trade: page 165.) Declining availability of 'wild' musk has led the Chinese and other nations to explore the possibility of maintaining musk deer in captivity, and removing the musk from the live animal. Flerov (1952) noted that 'it is not necessary to kill the animal in order to obtain the musk. To remove the musk from the bag it is only necessary to put a tube into the aperture, when a stream is excreted by slight pressure on the bag.' Musk-deer farms have been established in a number of places in China, but the production of musk from these farmed deer is poor, with perhaps only 14 g extracted from each male in a year.

The practice of bringing deer into farms in order to exercise more control over their management, or to have readier access to the commodity for which man wishes to exploit them, is not new. The idea of bringing deer into captivity, almost into domestication, is after all merely one stage further than the semi-domestication of reindeer practised by the Lapps, and deer have been maintained in captivity in Asia for centuries so that velvet antler could be more readily harvested and more abundantly available. Deer parks were also established widely in mediaeval Europe, sometimes as hunting parks, but more commonly to provide a ready source of fresh venison for the Lord of the Manor. The remains of great deer parks and compounds, with great retaining walls and ditches, may be found all over Europe. Gradually, however, the venison-producing function assumed lesser importance; the deer parks that graced many of the stately homes of Britain and Europe were maintained purely for aesthetic reasons — for the decorative appeal of a herd of majestic red deer or graceful fallow below the great house itself. Within England and Wales, a recent survey shows that over 200 such parks are still maintained, many still in private ownership (British Deer Society survey).

The wheel has, however, come full circle. Over the last 20 years or so there has once again been increasing interest in the farming of deer for economic gain: for the production of antler and other by-products for the East, or for the production of venison. Red deer, wapiti and fallow in particular are becoming exploited once more in this way, and lesser numbers of rusa, sambar and even chital are being farmed in Australia and on the Asian mainland. Deer-farming is big business, and is attracting increased interest in the current recession being experienced in conventional agriculture. Deer farms producing velvet antler or venison are now found throughout Europe (in Denmark, Germany and Austria, as well as in the UK) and Australasia (in both Australia and New Zealand). In 1986 there were more than 400,000 farmed deer in New Zealand, distributed among some 3,500 deer farms, and the export market from that country alone in meat and velvet exceeded 30 million NZ dollars (approximately £14.5 m) (New Zealand Deer Farmers Association figures). In the UK, 50 or 60 red or fallow deer farms are currently registered with the British Deer Farmers Association. The farming of deer has developed into a major industry, and with it has grown a wealth of knowledge on ecology, behaviour and farm management. This is not the place to review this extensive and rather specialist literature, but the interested reader is recommended to browse through Fennessy and Drew (1985).

This discussion of the exploitation of deer would be incomplete without noting a few other — perhaps eccentric? — ways in which Man has used deer or their products. Several species have been put to harness, although the reindeer, as its name reflects, has always been the principal one used for draught purposes. Whitehead (1972) notes that a single reindeer, as a beast of burden, can carry a load of about 40 kg in saddlebags, while on a sled it can draw a load of about 200 kg over snow, travelling up to 40 km a day. Lapps frequently use their reindeer in this way and may harness nine or ten deer in single file to their sleds. For travel over grass and heather a modified sled can be used, and for roadwork wheeled vehicles have been successfully adapted. In many parts of Siberia, reindeer carry mail and haul buses (Whitehead, 1972). The European moose has also been trained to harness as a draught animal, as have red deer. Whitehead records that in the 18th century George Walpole, Third Earl of Oxford, was known to drive around the Newmarket area (Suffolk, England) with four red stags drawing his phaeton in place of horses.

DEER AS PESTS

While deer populations may be seen by Man as a resource for exploitation or recreation, equally commonly they come to intrude upon his agricultural or other interests as pests. Deer at high densities can cause a dramatic impact on vegetation (Chapter 6 and, for example, Mark and Baylis, 1975; Ashby and Santiapillai, 1984). Having seen the effects of captive *Axis porcinus* on his zoo pastures, one head keeper quipped 'if the Almighty already knew about hog deer, why did He bother to inflict the East with plagues of locusts?' (Hall, 1987).

Deer may conflict with Man's interests in agriculture and silviculture; at unnaturally high densities they may threaten conservation of natural communities. Yet in all cases the conflict is in a sense artificial. Where deer grazing threatens native rangelands and the plant and other animal life associated with such communities, it is because the deer have reached unusually high densities in such areas. This in itself is usually the result of human encroachment, eroding deeper and deeper into the availability of natural habitats, forcing the deer themselves to concentrate on smaller and smaller patches; or — with fine irony — the deer causing the damage may not in any case be native species, but exotics introduced oringinally by Man himself. Rusa deer introduced into Papua New Guinea reached tremendous population levels and greatly threatened the stability of the natural grasslands and wetlands (Fraser Stewart, 1981), and we have already mentioned the degree of damage caused to the native vegetation of New Zealand by the introduction of deer in the 1900s (Holloway, 1950; Mark and Baylis, 1975; Challies, 1985).

The same factors of erosion of natural habitat or the population explosions of species introduced by Man also seem to lie behind many of those instances where deer become pests of Man's agricultural or timber crops. This is not a universal explanation, and in many cases natural species may assume pest status; even here, however, the situation is artificial. Man is changing the environment in cultivating his farm or forest crops, and is himself creating the conflict by replacing the natural vegetation upon which the deer normally rely with his own, man-made communities, often giving the deer little choice but to exploit these new environments. The fact that Man's agricultural crops tend to be highly nutritious may of course exacerbate the problem, increasing the carrying capacity of the environment and boosting the productivity of the deer populations themselves. Whatever the cause of the conflict, Man frequently finds himself in competition with deer; and the effects of deer upon a crop may be devastating.

Deer can cause damage to agricultural or forest crops in a number of ways. As herbivores, they may graze farmland or browse shoots, buds and leaves from young trees just in the practice of normal feeding behaviour. They may dig up root crops in winter or strip bark from the trunks of timber species in the same search for food. In the season, when the velvet dries on the new antler growth, males rub their antlers up and down on pliant saplings to clean them of the irritant dead skin, in the process shredding the trunk and branches of the young tree (Plate 22). Similar damage to young trees and other vegetation may be caused during aggressive display, particularly among territorial species. All these behaviours are quite normal but damage may also be caused by abnormal behaviours, or by normal behaviours performed with greater than normal frequency or intensity. The incidence of thrashing and fraying damage in European woodlands by roe bucks is greatly increased where, through Man's mismanagement, population densities are unrealistically high or population structure, in terms of age and sex distribution, has been distorted (Putman, 1985).

The damage caused may sometimes be extensive. In subsistence agricultural systems, grazing by deer may wipe out an entire crop in a single night. Even in more intensive agricultural systems, deer grazing may cause severe damage, while browsing or bark-stripping of forest trees, if it does not result in death of the affected trees, will severely check their growth. Precise figures are hard to come by, but Staines (1985) calculates that deer damage in commercial coniferous forests in Scotland (due primarily to red and roe deer) can be worth up to 40 per cent of the crop; Putman (1985) records that grazing by roe deer in cereal fields in southern England could affect up to 30 per cent of the crop. It is, however, important to note that this is probably exceptional and often, although the apparent damage may be severe, the actual economic loss which results may be far less significant.

While, for example, in Putman's study of damage by roe to cereal crops in southern England up to 30 per cent of the total area of a field might be affected, this was a maximum level rarely encountered. More commonly, less than 5 per cent of the crop might be grazed. Further, the timing of damage is critical and, by coincidence, the major period of grazing of cereal crops generally ended in mid to late April (when alternative food supplies begin to become more available in the woodland communities within the deers' range). If grazing does stop at this time, the crop may show remarkable powers of recovery. Each damaged plant puts on a compensatory spurt of growth, catching up with the growth stage of the rest of the crop. Grazing itself promotes tillering of the crop (the production of side shoots), so that after grazing there may in fact be *more* grain-bearing stems — although the individual ears are usually somewhat smaller. Owing to a combination of all these factors, despite extensive apparent damage early in the year, by harvest in late summer yields of grazed plots were not significantly lower than those of ungrazed areas in the crop (Table 8.1). Nor are these results atypical. Field roe of

Table 8.1 The effects of damage by roe deer on cereal yields. Plots of winter wheat and spring barley severely damaged by roe deer in late spring were harvested in July, one week before normal harvest. Yields of damaged plots are compared with those of undamaged, control, areas. Figures are presented as means with SD in brackets. No differences between damaged and undamaged plots are statistically significant (data from Putman, 1985)

	Spring barley (1983)		Winter wheat (1984)	
	Number of grain-bearing stems per m^2	Weight of unthreshed ears (grams per m^2)	Number of grain-bearing stems per m^2	Weight of unthreshed ears (grams per m^2)
Damaged by roe	787 (84.2)	493 (100.1)	330 (63.9)	780 (147.9)
Undamaged	672 (113.1)	578 (136.7)	367 (65.8)	720 (123.3)

central Europe (page 38) are entirely reliant on the agro-ecosystem for most of the year, yet, despite this, damage to field crops (maize, barley, rye, etc.) during vegetative growth is considered to reduce yield by less than 1 per cent (Kaluzinski, 1982; Obrtel and Holisova, 1983; Obrtel *et al.,* 1984).

Of course, this refers only to damage during the vegetative stages of crop growth; later damage, involving direct removal of ripening grain, will, if it occurs, cause measurable and permanent loss. In addition, deer can represent a serious pest problem to other types of agriculture: to market gardeners, nurserymen and horticulturists. In orchards, too, losses may be significant. In America, white-tailed deer are considered second only to birds in importance as orchard pests (Anthony and Fisher, 1977), reducing fruit production by killing trees, reducing vitality, or in direct removal of fruit buds (Boyce, 1950; Eadie, 1961; Beckwith and Stith, 1968; Katsma and Rusch, 1980).

Finally, as we have noted, deer may cause extensive damage to timber. Primarily woodland species by nature, they quickly establish resident populations even in newly afforested areas (resident populations of red deer were well established within the fences of commercial coniferous forests planted in previously open ground in Scotland within 15–20 years of planting: Ratcliffe and Staines, 1982). While those species which are preferential grazers may cause relatively little damage, browsing species can cause heavy losses. Browsing is of course most detrimental to young trees whose terminal buds or major branches are within reach of the deer, but even after they have grown beyond this stage they are still vulnerable to feeding damage through bark-stripping and browsing of lateral shoots, and to damage from antlers during velvet-cleaning or aggressive display. In areas of high deer density, death of trees or severe checking can, as we have noted, lead to an estimated 40 per cent loss of yield (Ratcliffe and Staines, 1982; Staines, 1985).

Ratcliffe and Staines's studies have been undertaken primarily within single-species plantings of Norway or Sitka spruce in Scotland, where damage is predominantly by red deer, but similar levels of damage are apparent in other systems. In a study of damage by roe deer in an area of mixed forestry in Scotland, Cousins (1987) noted that nearly all trees in coniferous plantations (tree heights <2m) and among blocks of young goat willow (*Salix caprea*) had suffered recent damage, with 20–30 per cent of trees damaged even in older conifer stands (of up to 5 m in height). Overall, 70 per cent of recorded damage was due to browsing (of leading shoots, 13 per cent; of both leading shoots and lateral branches, 22 per cent; of lateral branches only, 35 per cent), with 30 per cent contributed by antler damage.

Such severe damage occurs only in areas of high deer density, and, as in agricultural systems, the damage may at times be more apparent than real. Despite the extensive damage recorded in her study, Cousins was unable to detect any evidence of loss of yield. Of a total of 300 damaged trees tagged, only two actually died as a result of damage, and the growth rate of damaged Scots pine, Sitka spruce or larch was not measurably different from that of undamaged trees; for deciduous species, those trees suffering lateral damage actually grew faster than undamaged ones (Cousins, 1987). Staines and Welch (1984) also record that the effect of damage by red deer on growth of Sitka spruce in Scottish commercial forests is less than might be anticipated from the degree of damage experienced, since, where a terminal bud has been removed, a lateral shoot may assume dominance, moving up ('flagging') to take over at the crown of the growing tree. Despite such instances, however, there is no doubt that, as in agricultural systems, deer at sufficiently high density *can* inflict considerable damage of real economic significance.

Wherever deer are implicated as pests, the response is usually simple and predictable: an attempt to reduce population numbers of the pest. In fact there are a number of problems with adopting such an obvious approach. Reduction in numbers, by whatever method, of the pest species concerned may seem a clear and logical solution to contain any pest problem, but it is biologically somewhat naive. Any reduction in numbers achieved, short of total extermination, is short-lived. Control effort must be repeated year after year, for the natural response of most animal populations to a decrease in their numbers is an increase in productivity: increased recruitment through both reproduction and immigration and increased survival. Further, any culling programme is bound to disturb the population, and it may actually exacerbate the very problem it aims to contain: (i) by dispersing the pest from the initial area of control into new areas perhaps up to that time relatively free from damage; and (ii) by changing the behaviour of the individual animals themselves, making them more secretive and elusive and thus more difficult to 'contact' in any future cull. Frequently, population reduction is in practice the only realistic way of relieving an immediate pest problem, but we shall review in the concluding pages of this chapter a number of possible alternatives.

THREATENED SPECIES AND CONSERVATION

Where exploitation by Man, or his attempts at population control, are unrestricted or ill-conceived, human pressure may sometimes severely threaten the survival of the species. Over-exploitation of wild musk deer for the musk pods of the males has led to a dramatic decline in their numbers throughout their range. On the basis of known exports of the musk itself, Green (1985) calculates that between 20 per cent and 50 per cent of the entire population of Himalayan musk deer must have been killed each year between 1974 and 1980, and there is no doubt but that this had a tremendous impact upon that species, whose numbers are now estimated at some 30,000 animals.

Excessive and unrestrained hunting for meat, or for recreation — indeed for whatever commodity — may also cause dramatic declines in population numbers. While white-tailed deer are today widespread throughout North America, and resident in practically every state of the Union except Alaska and possibly Utah, this is in fact a relatively recent return towards their original status. Numbers within the US were estimated by Whitehead in 1972 at approximately 8 million deer (no figures are available for Canada, Mexico or Central America). Estimates of populations present at the beginning of the 19th century (Seton, 1910, in Whitehead, 1972) were in the region of 40 million, but over the years the deer were so heavily persecuted that by the end of that century the stock had reached a very low ebb and in some states had been virtually exterminated. During the early years of the present century a number of refuges were established and deer were reintroduced into areas from which they had been eliminated. In consequence, populations have

Table 8.2 Geographical location and current conservation status of the threatened deer of the world. (Based on data from Cowan and Holloway, 1978, updated where more recent information is available)

(a) Endangered taxa	Distribution and status	Notes
Muntiacus feae: Fea's muntjac	Restricted distribution in mountainous evergreen forests of Burma and Thailand. Numbers unknown.	Decline due to habitat erosion and forest clearance, exacerbated by hunting by local peoples.
Cervus duvauceli duvauceli: northern swamp deer	Nepal and northern India in grassy swamps and dry grasslands. Numbers estimated at 41,500.	60% of total world population occurs in two protected areas: Sukla Phanta Reserve in Nepal and Dudhura National Park (India). Principal threats habitat erosion through stock grazing and grass-cutting.
Cervus duvauceli branderi: southern swamp deer	Central India, associated more with drier forest grassland type than the northern race. Numbers declined from 3,000 in 1938 to 66 in 1969; in 1978 estimated 283.	Occurs only in one 940-km^2 area: the Kanha National Park in Central India.
Cervus eldi eldi: Manipur brow-antlered deer or sangai	Confined to one national park in Manipur State, India. The most threatened deer in the world, numbers in 1977 estimated at 18.	One of the three subspecies of *Cervus eldi,* two of which are endangered.
Cervus eldi siamensis: Thailand brow-antlered deer	Open plains and deciduous forests in Laos, Cambodia, Vietnam and Kainan Island. Population unknown.	Populations seriously depleted by uncontrolled hunting and destruction of habitat.
Cervus nippon taiouanus: Formosa sika	Formerly widely distributed in grasslands and open woodlands of Taiwan. Numbers unknown, possibly extinct in the wild.	Populations decimated by uncontrolled hunting and loss of habitat to cultivation. Thought to be extinct in the wild since 1969, but is maintained in captivity and reintroduction programmes are being considered.
Cervus nippon mandarinus *Cervus nippon grassianus* *Cervus nippon kopschi*	Occur only in a limited area within China in forest areas. May be extinct in the wild.	Populations depleted by over-hunting and loss of forest habitat to agriculture.
Cervus elaphus bactrianus: Bactrian or Bokharan deer	Riverine woodland in arid regions of parts of USSR and northern Afghanistan. Numbers estimated 1978 as >1,000.	Population has doubled since mid 1960s owing to relocation into protected areas.
Cervus elaphus barbarus: Barbary deer	Confined to a restricted area of cork-oak and pine forest on the border of Algeria and Tunisia. Total population size, including 150 in captivity: 500 animals.	
Cervus elaphus corsicanus: Corsican red deer	Confined to three mountainous areas in southern Sardinia and considered extinct in Corsica. Population estimated at c.200.	Numbers reduced by hunting, poaching and habitat degradation.
Cervus elaphus hanglu: hangul or Kashmir deer	Occurs only in moist temperate forest, in the vale of Kashmir. Estimated 250 individuals.	Numbers drastically reduced by poaching, loss of habitat to agriculture and human disturbance.
Cervus elaphus wallichi: Shou	Temperate forests in southeast Tibet and adjacent Bhutan. Believed extinct.	

Cervus elaphus yarkandensis: Yarkand deer	Riverine woodland in Sinkiang district, Chinese Turkestan. Status unknown, possibly extinct.	Heavily hunted; large areas of habitat lost to agriculture.
Cervus canadensis macneilli: McNeill's deer	Occurs in high-altitude forest — rhododendron areas of Sinkiang, Chinese Turkestan. Current status uncertain.	Populations threatened through overhunting for antlers for medicine trade. Now protected.
Dama dama mesopotamica: Persian fallow deer	In the wild, restricted to two areas of floodplain forest in western Iran, with wild populations estimated at 45–65 animals.	The animal is maintained in captivity, with two viable captive-breeding units.
Odocoileus virginianus leucurus: Columbian white-tailed deer	Floodplain forests of lower Columbia river in Washington and Oregon states. Numbers estimated at 400 animals.	
Odocoileus hemionus cerrosensis: Cedros Island deer	Confined to an area of pine forest and chaparral in one part of Cedros Island off Baja California. Estimated 50 individuals.	
Ozotocerus bezoarticus celer: Argentianian pampas deer	Currently occurs in only four locations in the Buenos Aires and San Luis provinces of northern Argentina, and numbers are estimated at around 200.	Decline attributed to over-exploitation by hunters and loss of habitat to agricultural development. This is a species of open grasslands and is further threatened by competition with domestic livestock grazed in native range and disease transmitted from the domestic stock.
(b) Vulnerable taxa		
Axis calamianensis: Calamian deer	Occurs only on a few islets in the Calamian Islands of the Philippines. Total population estimated at 900 animals and declining.	Initial reduction in numbers due to uncontrolled hunting; now threatened by loss of habitat owing to shifting cultivation.
Hippocamelus antisensis: Peruvian huemul	High-altitude forests and scrub of the Northern Andes. Total population size unknown, but considered rare over most of its range except in Peru, where it is locally numerous.	Populations low and believed to be still declining owing to poaching, loss of habitat to agriculture at lower altitudes, and competition with domestic stock at higher altitudes.
Blastocerus dichotomus: marsh deer	Seasonally-flooded grasslands along rivers in southeast Peru, Bolivia, Brazil, Paraguay and northern Argentina. Estimated at c. 6,000 animals.	Numbers declining owing to poaching, habitat encroachment, and disease transmitted by domestic stock.
Moschus m. moschiferus: Himalayan musk deer	High-altitude woodland and shrub in Himalayas from northern Pakistan to Burmese border. Numbers perhaps 25,000–30,000	Populations have been depleted primarily through hunting, for musk.
(c) Taxa regarded as rare or of indeterminate status		
Muntiacus crinifrons: black or hairy-fronted muntjac	Recorded only from the state of Chekiang in southeast China. Status unknown.	
Axis kuhli: Kuhl's deer	Confined to Bawean Island (220 km^2) in Indonesia. Total population estimated at 200–500 animals.	Numbers are increasing now that hunting has been controlled. Main threat now is loss of habitat to cultivation and teak plantings.
Cervus albirostris: Thorold's deer	Occurs in high-altitude coniferous forest and rhododendron in eastern Tibet and China. Numbers unknown.	Seriously depleted by hunting for antlers for medicine trade.
Hippocamelus bisulcus: Chilean huemul	High-altitude forest and shrub of Andes in southern Chile and Argentina. Numbers uncertain, between 400 and 900.	Numbers believed to be declining as a result of poaching, disease, and loss of habitat due to fire, felling, grazing of domestic stock.

gradually built up once more; even today, however, these are but a fifth of their former numbers.

Data on populations of wapiti in Yellowstone National Park in Wyoming (Houston, 1982) offer in microcosm a similar illustration of the potential influence of human predation on deer populations, although in this case we are considering changes in single local population. Houston suggests that the same intense hunting pressure in the late 19th century that so dramatically affected white-tailed deer over the entire region must have reduced numbers of wapiti in the Yellowstone area (9,000 m^2) to an all-time low by the mid 1880s. Hunting in the Park was prohibited in 1883 and although, owing to the difficulty of enforcing the legal protection, recovery was initially somewhat slow, by the end of the 1920s populations had grown to an estimated 12,000–16,000 animals: a figure considered by Houston to be the Park's natural capacity for the deer. Licensed hunting was once more permitted within the Park, and hunting pressure increased during the period from the 1930s until 1968 (until a moratorium on all hunting was imposed in 1969). Changes in the population over this period offer a perfect carbon copy of the situation apparent in the late 19th century and numbers of wapiti in the Park fell again from around 12,000 to around 4,000 (Houston, 1982): demonstrating only too clearly the consequences of over-exploitation.

It is perhaps unusual for hunting, on its own, to result in such severe population decline; after all, we noted above that attempts to exterminate pest species were likely to prove unsuccessful! More commonly, population decline of this type is brought about by hunting in conjunction with other pressures: habitat erosion or competition with domestic livestock.

The fact remains that, while only one species of deer (*Cervus schomburgki,* last recorded in Thailand in 1932) has become extinct in the last century, of the 40 species alive today, no fewer than 29 subspecies are listed by the IUCN as endangered or vulnerable; among that list of 29 are included 12 full species (Cowan and Holloway, 1978). On this list, the Himalayan musk deer is classed as vulnerable (defined as taxa of which most or all populations are decreasing, and those with populations that have in the past been seriously depleted and whose ultimate security is not yet assured). In the same category are the Calamian deer, the Peruvian huemul and the marsh deer of South America. Considered endangered (in danger of extinction within the near future, often with populations as low as 500 individuals) are Fea's muntjac, the Persian fallow deer (*Dama dama mesopotamica*), both subspecies of swamp deer or barasingha and two subspecies of the brow-antlered deer. A number of distinct subspecies or geographical races of *Cervus elaphus* and *Cervus nippon* are also threatened, along with particular subspecies of white-tailed and mule deer. Finally, all subspecies of the South American pampas deer are considered rare, and the Argentinian form *Ozotocerus bezoarticus celer* is classed as endangered. (For a summary of the distribution of all these animals, and their status in 1977, see Table 8.2.) The majority of these deer have suffered considerable reductions in numbers from over-exploitation; all have been affected, to varying degrees, by modification of their habitat. Losses of habitat to agriculture, especially, have been particularly severe for those species which occur in moist, lowland areas (swamps, floodplains or riverine forest) within fairly arid regions (Cowan and Holloway, 1978).

MANAGEMENT FOR THE FUTURE

We began this chapter with the comment that few populations of deer in the world are today not influenced by Man, and it is clear that Man must assume responsibility for future management of the world's deer. Careful management must be directed towards ensuring sustained harvest from those species to be exploited, effective control of those species which are considered pests, planned conservation of endangered or threatened species, and also towards ensuring that populations even of the more widespread and abundant species are maintained in balance with the carrying capacity of the environment.

In the past, such attempts at management have not been universally successful: those responsible perhaps did not appreciate the need for management in the first place, or whatever management was imposed was ill-conceived. Indeed, the historical record is a rather sorry catalogue of over-harvesting of exploited populations, blind to the long-term consequences, and lamentable failure to achieve control of pest populations.

Management for Exploitation

Only too frequently in the past, those exploiting deer populations for meat or medicines took what they could while they could, unaware of — or oblivious to — the fact that the resource itself, while self-renewing, was not inexhaustible. We have noted 29 subspecies of deer now threatened or vulnerable, most as a direct consequence of over-exploitation in the past. A further two subspecies of wapiti, and single full species, *Cervus schomburgki*, have actually become extinct in the last hundred years or so, and many other species, while perhaps never driven to the brink of extinction, certainly suffered a tremendous decline in abundance through over-exploitation (see above). The tragedy is that in most cases exploitation can actually enhance population performance and, if carefully conceived, can return a regular and sustainable yield from the harvested population whereas the short-sighted overkill of the past could yield only short-term profit for a very limited period as populations declined.

We noted in Chapter 5 that most deer populations appear to respond to increasing density by a reduction in fecundity and an increase in mortality (particularly the mortality of juveniles withstanding their first harsh season of shortage: the temperate winter or the tropical dry season). Ultimately rates of reproduction and rates of mortality or emigration will reach a balance, so that the net rate of increase within the population becomes zero and the population numbers stabilise at some equilibrium level determined by the availability of environmental resources. Within such populations, exploitation acts to reduce numbers. In so doing, it brings the population once more below the level at which it is limited by environmental resources: effectively releasing the density-dependent brake on population growth. Reproduction increases, juvenile mortality falls, and the whole population age structure shifts towards the young animals, which have a faster growth rate and higher efficiency of food conversion. Productivity of the population rises.

This increased productivity of a population under exploitation is now well documented (e.g. Watt, 1955; Silliman and Gutsell, 1958; Gulland,

1962; and see also Eltringham, 1984). The exploitation, by reducing density, lessens competition and enhances productivity: *producing* the surplus that is then harvested, and producing thus a sustainable yield. Exploitation is now creaming off interest on the capital, rather than eroding the capital itself.

The increase in productivity is, however, limited. It will occur only if (a) the species exploited is one which has the capacity to respond to the changes in its own density with changes in reproduction and/or mortality rates, and if (b) the population is at the time limited by availability of some resource and its potential productivity is thus already suppressed. (Clearly, a population well below its environmental ceiling is already exercising its maximum rate of increase and a further decrease in numbers cannot boost production further.) In the same way, the increase in productivity which is achieved under exploitation can continue only up to that level at which the density-dependent brake operating on population growth is fully released and the population reaches its maximum rate of increase once more. Nonetheless, for populations which do respond to density in this way, exploitation up to that maximum level should increase productivity, in theory and in practice, and offer to the exploiter a long-term sustainable harvest. So long as we can calculate what is the level of exploitation at which a population will return to its maximum potential rate of increase, then exploitation poses no threat and arguably might even be considered an active form of conservation of the exploited species both through its influence in stimulating population performance and on more prosaic grounds of economics. Careful management is more likely to be assured for a species which represents a significant source of income to some human society than for some other animal of no commercial value.

Incidentally, while in theory exploitation will increase productivity in a population right up to that point where the population reaches its maximum potential rate of increase, it does not necessarily represent the level at which exploitation is at its most efficient. Productivity of the deer population may be at its maximum, but the yield may not. As Slobodkin (1968) pointed out, for natural predators:

> The process of predation increases the rate of 'manufacture' per unit of prey population but decreases the size of the prey population. An extremely small prey population cannot produce the same total quantity of yield to the predator as a somewhat larger one, even if the large one is producing at a slower rate per animal.

Such considerations lead to a redefinition of optimal rates of exploitation, not as those which maximise productivity of the prey but as those which produce maximum sustained yield to the exploiter.

There is one further subtlety. If exploitation is selective, removing preferentially those individuals of least use to the exploited population (post-reproductive adults, surplus young, surplus males in harem species), optimal levels of exploitation and yields may well be higher than in those cases in which the human predator acts in a random fashion, for in the former case predation is reducing overall numbers and intensity of competition without reducing the size of the reproductive element of the population. Further, by removing from the population those animals which are the ones most likely to die anyway, the exploitation is altering the natural pattern of mortality as little as possible and is thus less likely to upset the overall equilibrium of that population (Slobodkin, 1968) and risk its extinction. Much research is now being directed at establishing those exploitation levels that realise maximum sustained yield from different populations of large animals, including deer (e.g. Beddington, 1973), so that management for the future may be more biologically realistic, and populations conserved.

Management for Control

This rapid response of populations to man-imposed changes in their density also explains in part why direct population reduction is unlikely, on its own, to prove effective as a solution to pest problems. Attempts at population reduction once again merely serve to boost productivity, and this means that any control policy based on population reduction must be sustained and repeated year after year: both financially and in terms of manpower a costly decision. Quite simply, many populations which have reached a density at which they constitute a problem as a pest are probably suffering reduction in recruitment and survival due to the effects of that same density. If population levels are lowered artificially, the density-dependent brake on population growth is released: reproduction increases, mortality declines. Even in populations where we cannot prove such density-dependent effects on recruitment and survival, local reduction of the species in one area is rapidly compensated for by immigration from outside.

Population reduction is thus a perpetual treadmill. *If carefully carried out*, it may offer short-term (and quick-acting) relief of pest pressure in particularly badly affected areas. As a long-term solution it is an extremely expensive proposition to contemplate. Further, all this has assumed a population reduction intelligently planned. Indiscriminate culling, or a cull concentrating on the wrong age- or sex-class of animal, can prove not only expensive but counterproductive. We have already noted (page 165) that the disturbance caused by trapping or shooting may result in fragmentation and dispersal of the population to new areas, or may cause a shyness of behaviour which makes future control even more difficult. A clear illustration of this last point may be taken from consideration of the red deer populations in New Zealand.

The introduction of red deer to New Zealand and their subsequent rapid increase to pest proportions is well documented (Challies, 1985; see also pages 120, 154–5). Government-promoted control schemes were introduced in 1931 and have continued in various forms to the present day. Control operations initially involved hunting on foot, with or without dogs. Although high cull figures were achieved, numbers of deer killed per man remained constant over many years, suggesting that the hunting had little real impact on deer numbers. The introduction of hunting by helicopter improved access to more inaccessible terrain and, with the establishment of a new deer-farming industry in the early 1970s, such hunting intensified as the demand for live animals suddenly made it extremely lucrative. Helicopter crews could catch or kill five or six deer per flying hour (up 12,000 deer per helicopter year). The impact was colossal, but restricted largely to open areas on hill tops or screes. Numbers appeared to decline dramatically, to a point where, in 1982, many helicopter crews went out of business because catch rates no longer justified the expense of the operation (Challies, 1977; 1985). Although there can be no doubt that numbers of red deer were indeed greatly reduced, it is unlikely that this alone contributed to the sudden decline in catch rate. Pellet-group counts on open range suggest deer densities only slightly lower than those of the middle 1970s, and it has been proposed that the real change was behavioural in that animals now positively avoid the habitats where they are more exposed to attack (Figure 8.3). If this is true, further culling of these feral populations becomes increasingly problematical.

Figure 8.3 The change in density and, most importantly, in distribution of red deer in Westland, New Zealand, between 1969 and 1976. Figures, based on the density of dung-pellet groups recorded, illustrate the effect that intensive helicopter hunting, mainly above the timberline, has had on deer density and distribution. Source: Challies (1985)

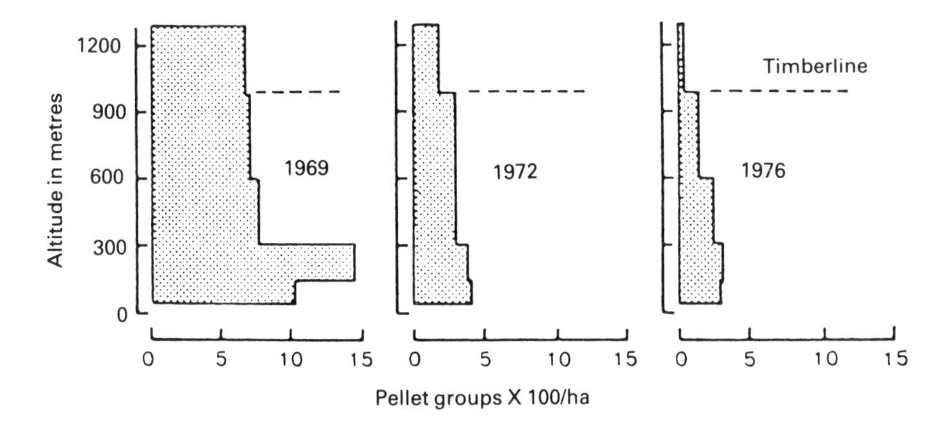

In addition to such general effects on behaviour, lack of selectivity in the cull, or selection of the wrong age- or sex-classes, may also exacerbate the pest problem by disrupting the social structure and social organisation of the pest. This may have two consequences: a further increase in population density, and an increase in damage caused because the animals display abnormal behaviour in response to a distorted social structure. We may take here a well-worn example of the effects of uninformed attempts at management of roe deer, an example the more pleasing because its lessons have now been learnt by most managers. Before management policies were modified to take account of an increasing awareness of the animal's social behaviour, the strategy adopted in controlling pest populations of roe in timber or farmland was to shoot as many as possible or as considered necessary. The stalkers shot on sight. Yet roe deer are territorial; the most obvious individuals and thus those most regularly encountered are territorial bucks. This, coupled with a preference on the part of the stalker for trophy heads, resulted in a selective cull of adult males. As a result, many territories were left vacant. Young bucks competed among themselves to take them over (causing considerable incidental damage in display) and, since younger individuals may be satisfied with smaller territories, effective density overall *increased*. By leaving the territorial bucks alone (as is now normal policy, in recognition of this behavioural fact) a constant density is maintained, as sub-adults fail to establish territory and are forced to disperse from the area.

Overall, we may conclude that pest control through direct population control is at best costly, at worst counterproductive. But what are the alternatives? Perhaps the first and easiest option is *laissez-faire*. In many cases an alleged pest problem may prove on closer examination to be insignificant, or, at least in economic terms, less costly to accept than to prevent. Adoption of even effective population control measures may prove more expensive than the loss sustained; risks of increasing damage through ill-considered or ineffective control measures may outweigh the costs of accepting current damage levels. We noted earlier for example that, although roe deer in southern England and elsewhere in Europe may appear to cause extensive damage to cereal crops through grazing and trampling, so long as this damage is restricted to the vegetative phase of growth, the plants are capable of considerable compensatory growth, so that by harvest the actual *economic* significance of such damage may be much reduced.

Clearly, the timing of pest impact in relation to the growth pattern of the target crop is going to have a profound influence on the potential recovery, and will thus greatly affect the degree of economic loss. Yet in the cases described, despite the dramatic appearance of the apparent damage when first sustained, actual economic loss was insignificant. The pest problem was illusory and any costs of management would have been wasted. In other cases economic loss may be real, but still less costly than the expense of control measures required to reduce that loss.

It is normal policy in commercial softwood forestry programmes to plant the fastest-growing species compatible with soil and climate of the area. Yet such species are frequently very susceptible to damage by deer browsing. The economic loss sustained may be considerable, either through direct damage or through necessary costs of fencing and deer-culling. The operation in total may prove less cost-effective than planting of timber species less palatable to deer or more tolerant to browse damage. Such species in general do not show such rapid growth rates as the more favoured commercial softwoods and time required before harvest may be increased, reducing apparent profit; nonetheless, through reduced costs of deer damage or deer control they may prove overall a more economic alternative.

These are, of course, isolated examples, and often some form of animal management may be justified. Even if the decision is taken that some form of management is required, actual population reduction, however, is still not necessarily the obvious or most cost-effective solution. If, by careful research, we can establish why an animal follows a particular behaviour pattern which causes it to conflict with Man's interest, we may be able to manipulate the environment so that it performs that behaviour elsewhere, or no longer needs to perform it. Alternatively, we may be able to modify the behaviour directly, or even exploit our knowledge of the animal's whole behavioural repertoire in order to minimise its pest status.

We have already noted that the use of agricultural crops by roe deer is highly seasonal. Roe deer are well known to be 'concentrate selectors'

(Chapter 3), needing to feed selectively on small morsels of highly nutritious foodstuffs. Close examination reveals that their use of agricultural pastures and cereals corresponds very closely to a time when these are the only vegetational communities showing new growth (and thus offering maximum digestible nutrients). As soon as the grasses of woodland glades or trackways begin to flush, or buds burst on woody browse species, the animals switch their attention to these preferred foodstuffs and stop feeding on crops. Growth of winter cereals or fodder crops such as turnips or kale in the autumn and winter likewise coincides with the end of the growth period in the more 'natural' communities. Should we feel a need to manage roe deer away from arable crops, therefore, we might achieve this most successfully by some form of habitat alteration which would offer highly nutritious forage elsewhere in the animal's range at these critical times of year. If offered in equivalent quality but without the drawback of the necessary exposure experienced by deer feeding out in open field, such 'honeypots' might well be preferred, and use of farmland would be correspondingly reduced.

Grassed clearings in woodland, artificially sown and artificially fertilised, were once an important feature of deer management throughout Europe, drawing deer to central 'lawns' where they could be counted more easily, and more easily shot, as well as helping to keep them off open ground surrounding the deer forest. The establishment in commercial woodland of an understorey, of bramble and rose, hawthorn, blackthorn, willow and hazel, might also reduce the need for foraging outside the woodland, or on the commercial timber crop (while at the same time permitting an actual increase in deer density). Even the use of artificial supplements may prove valuable. Mineral licks may prove effective in holding deer or in reducing certain types of damage such as bark-stripping (de Nahlik, 1959) when such damage can be shown to be due to dietary deficiency. It has been claimed more recently that sugar blocks may also be effective in reducing bark-stripping (Fischer, 1975). Deer in a German park were provided with mineral and sugar licks as routine. Under this regime, no bark-stripping damage was experienced within the deer enclosures. When sugar blocks were withdrawn, heavy bark-stripping was recorded; damage ceased immediately when sugar licks were redistributed.

It is to be hoped that gradually such habitat manipulation will come back into vogue for deer and other vertebrate pests. With increasing understanding of the natural behaviour and ecological requirements of our pest species, there is certainly potential for successful management in this way. It has the signal advantage that, once effective, it is permanent.

Management for Conservation

In our discussions of management, for control or exploitation, we have repeatedly stressed that most natural populations respond to reduction in numbers by increased productivity. Yet there is no doubt that there is a tragically long list of species and subspecies of deer whose numbers have been woefully reduced (see Table 8.2), apparently through over-exploitation. It is important to remind ourselves in this context that enhanced production within a population suffering some degree of imposed mortality will occur only if the population is capable of responding to changes in density with changes in reproduction and mortality, and in addition only if it is, not already so far below its environmental ceiling that it is already reproducing at its maximum rate (page 170). Once a population has dropped below that critical level, continued exploitation can act very quickly to cause a drastic decline in numbers; and many factors can combine with heavy exploitation to cause the initial drop below that threshold. We noted on p. 166 that while most of the taxa currently threatened have indeed suffered considerably from over-exploitation, this has also been accompanied in every case by modification of their habitat or by actual habitat loss.

Indeed, habitat erosion on its own can also result in population decline, even where it is not accompanied by heavy exploitation, if the species concerned is too 'conservative' in its ecology, too much of a habitat specialist. Animals which have adapted specifically to a particular, highly specialised set of resources are likely to be far more sensitive to habitat change, and cannot compensate for habitat loss by making greater use of alternative habitat types. It is the opportunists that survive — it is in fact the opportunists that tend to be pests!

Here again, future management must combine changes in management of the populations themselves with habitat manipulation — controlling continuing over-exploitation, halting habitat erosion, perhaps even re-creating suitable habitat, and subsidising natural populations with reintroductions from captive stock. It is to be hoped that it is not too late.

References

Albon, S.D., Mitchell, B., and Staines, B.W. (1983) 'Fertility and body weight in female red deer: a density dependent relationship', *J. Anim. Ecol.* **52**, 969–80.

Albon, S.D., Clutton-Brock, T.H., and Guinness, F.E. (1987) 'Early development and population dynamics in red deer. II Density-independent effects and cohort variation', *J. Anim. Ecol.* **56**, 69–82.

Allen, A. (1983) 'What the deer sees', *The Field*, 12 March, 1983, p. 419.

Andersen, J. (1961) 'Biology and management of Roe-deer in Denmark', *La Terre et la Vie* **108**, 41–53.

Anderson, A.E. (1981) 'Morphological and physiological characteristics'. In: *Mule and Black-tailed Deer of North America*, ed. Wallmo, O.C., pp. 27–97. University of Nebraska Press.

Anderson, C.C. (1958) 'The elk of Jackson Hole', *Wyoming Game & Fish. Comm. Bull.* **10**, 1–184.

Anthony, R.G., and Fisher, A.B. (1977) 'Wildlife damage in orchards – a need for better management', *Wildlife Soc. Bull.* **5**, 107–12.

Arman, P., Kay, R.N.B., Goodall, E.D., and Sharman, G.A.M. (1974) 'The composition and yield of milk from captive red deer (*Cervus elaphus* L.)', *J. Reprod. Fertility* **37**, 67–84.

Ashby, K.R., and Santiapillai, C. (1984) 'Deterioration of grazing associated with high densities of herbivores in Ruhuna National Park, Sri Lanka', *Acta Zool. Fennica* **172**, 257–8.

—— (1986) 'The life expectancy of wild artiodactyl herbivores: water buffalo (*Bubalus bubalis*), sambar (*Cervus unicolor*), spotted deer (*Axis axis*) and wild pig (*Sus scrofa*) in Ruhuna National Park, Sri Lanka, and the consequences for management', *Tigerpaper* (FAO) **13**, 1–7.

Baker, K. (1973) Reproductive biology of fallow deer (*Dama dama*) in the Blue Mountains of New Zealand. MSc thesis, University of Otago, N.Z.

Baker, R.H. (1984) 'Origin, classification and distribution of white-tailed deer'. In: *White-tailed Deer: Ecology and Management*, ed. Halls, L.K., pp. 1–18. Stackpole Books, Pennsylvania.

Balasubramaniam, S., Santiapillai, C., and Chambers, M.R. (1980) 'Seasonal shifts in the pattern of habitat utilization by the spotted deer (*Axis axis* Erxleben, 1777) in the Ruhuna National Park, Sri Lanka', *Spixiana* **3**, 157–66.

Beasom, S.L. (1974) 'Relationships between predator removal and white-tailed deer net productivity', *J. Wildl. Mgmt* **38**, 854–9.

Beckwith, S.L., and Stith, L.G. (1968) Deer damage to citrus groves in South Florida', *Proc. Ann. Conf. Southeast Assoc. Game and Fish Comm.* **21**, 32–8.

Beddington, J. (1973) The exploitation of red deer (*Cervus elaphus*). PhD thesis, University of Edinburgh.

Bell, P.J. (1876) 'Notes on the myology of the limbs of *Moschus moschiferus*', *Proc. Zool. Soc. Lond.* 182–8.

Bergerud, A.T. (1971) 'The population dynamics of Newfoundland caribou', *Wildlife Monographs* **25**, 1–55.

—— (1973) 'Movement and rutting behaviour of caribou (*Rangifer tarandus*) at Mount Alberta, Quebec', *Can. Field Nat.* **87**, 357–69.

—— (1974) 'Rutting behaviour of Newfoundland caribou'. In: *The Behaviour of Ungulates and its Relation to Management*, eds. Geist, V., and Walther, F., pp. 395–435. IUCN, Switzerland.

—— (1978) 'Caribou'. In: *Big Game of North America*, eds. Schmidt, J.L., and Gilbert, D.L., pp. 83–101. Stackpole Books, Pennsylvania.

Berwick, S.M. (1974) The community of wild ruminants in the Gir forest ecosystem, India. PhD thesis, Yale University.

Bildstein, K.L. (1983) 'Why white-tailed deer flag their tails', *Am. Nat.* **121**, 709–15.

Bobek, B., Perzanowski, K., Siwanowicz, J., and Zielinski, J. (1979) 'Deer pressure on forage in a deciduous forest', *Oikos* **32**, 373–9.

Boddington, H.J. (1985) The behaviour of fallow deer: time-budgeting with regard to animal class in a park herd of fallow deer (*Dama dama*). BSc Hons thesis (Biology), University of Southampton.

Borg, K. (1970) 'On mortality and reproduction of roe deer in Sweden during the period 1948–1969', *Viltrevy* **7**, 121–49.

Boyce, A.P. (1950) 'Orchard damage', *Michigan Conservation* **19**, 9–10, 26.

Boyce, M.S., and Hayden-Wing, L.D. (1979) *North American Elk: Ecology, Behaviour and Management.* University of Wyoming Press.

Bramley, P.S. (1970) 'Territoriality and reproductive behaviour of roe deer', *J. Reprod. Fert.* Suppl. **11**, 43–70.

Branan, W.V., and Marchinton, R.L. (1985) 'Biology of red brocket deer in Surinam with emphasis on management potential'. In: *Biology of Deer Production*, eds. Fennessy, P.F., and Drew, K.R., pp. 41–4. Royal Society of New Zealand.

Brander, A. (1923) *Wild Animals in Central India.* London.

—— (1932) 'The Central Provinces (Madhya Pradesh)'. In: *The Preservation of Wildlife in India*, ed. Burton, R., pp. 46–51. Bangalore.

Bresinski, W. (1982) 'Grouping tendencies in roe deer under agrocenosis conditions', *Acta Theriologica* **27**, 427–47.

Brooke, V. (1878) 'On the classification of the Cervidae, with a synopsis of the existing species', *Proc Zool. Soc. Lond.* (1878) 883–928.

Bubenik, A.B. (1965) 'Beitrag sur Geburtskunde und zu den Mutterkind-Beziehungen des Reh- (*Capreolus capreolus* L.) und Rotwildes (*Cervus elaphus* L.)', *Z. Säugetierk.* **30**, 65–228.

—— (1968) 'The significance of antlers in the social life of the Cervidae', *Deer* **1**, 208–14.

Bubenik, G.A., Schams, D., and Leatherland, J.F. (1985) 'Seasonal rhythms of prolactin and its role in the antler cycle of white-tailed deer'. In: *Biology of Deer Production*, eds. Fennessy, P.F., and Drew, K.R., pp. 257–62. Royal Society of New Zealand.

Carbyn, L.N. (1974) *Wolf Predation and Behavioural Interactions with Elk and Other Ungulates in an Area of High Prey Density.* Canadian Wildlife Service.

Caro, T.M. (1986) 'The functions of stotting in Thompson's gazelles: some tests of the predictions', *Animal Behaviour* **34**, 663–84.

Caughley, G. (1970) 'Eruption of ungulate populations with emphasis on Himalayan thar in New Zealand', *Ecology* **51**, 53–72.

Chadwick, M.J. (1960) '*Nardus stricta* L', *J. Ecol.* **48**, 255–67.

Chaetum, E.L., and Severinghaus, C.W. (1950) 'Variations in fertility of white-tailed deer related to range conditions', *Trans. N. Am. Wildl. Conf.* **15**, 170–89.

Challies, C.N. (1977) 'Effects of commercial hunting on red deer densities in the Arawata Valley, South Westland, 1972–1976', *N.Z.J. Forestry Science* **7**, 263–73.

—— (1985) 'Establishment, control and commercial exploitation of wild deer in New Zealand'. In: *Biology of Deer Production*, eds. Fennessy, P.F., and Drew, K.R., pp. 23–36. Royal Society of New Zealand.

Chapman, D.I. (1975) 'Antlers – bones of contention', *Mammal Review* **5**, 121–72.

—— (1981) 'Antler structure and function – a hypothesis', *J. Biomech.* **14**, 195–7.

Chapman, D.I., and Chapman, N.G. (1975) *Fallow Deer: their history, distribution, and biology.* Terence Dalton, Lavenham.

Chapman, D.I. Chapman, N.G., and Dansie, O. (1984) 'The periods of conception and parturition in feral Reeves' muntjac (*Muntiacus reevesi*) in southern England, based upon age of juvenile animals', *J. Zool. Lond.* **204**, 575–8.

Chapman, N.G. (1988) 'Muntjac'. In: *The Handbook of British Mammals* (3rd edn), eds. Corbet, G.B., and Harris, S. Blackwell, Oxford.

Chapman, N.G., and Chapman, D.I. (1980) 'The distribution of fallow deer: a worldwide review', *Mammal Review* **10**, 61–138.

Chapman, N.G., and Putman, R.J. (1988) 'Fallow deer' In: *The Handbook of British Mammals* (3rd edn), eds. Corbet, G.B., and Harris, S. Blackwell, Oxford.
Chase, W.W., and Jenkins, D.H. (1962) 'Productivity of the George Reserve Deer Herd', *Proc. Nat. White-tailed deer Disease Symp.* **1**, 78–88.
Chopra, I.C., Handa, K.L., and Kapur, L.D. (1958) *Indigenous Drugs of India.* U.N. Dhur, Calcutta.
Clutton-Brock, T.H. (1982) 'The functions of antlers', *Behaviour* **79**, 108-25.
Clutton-Brock, T.H., and Albon, S.D. (1979) 'The roaring of red deer and the evolution of honest advertisement', *Behaviour* **69**, 145–70.
—— (1982) 'Parental investment and the sex ratio of progeny in mammals'. In: *Current Problems in Sociobiology*, ed. King's College Sociobiology Group, pp. 223–48. Cambridge University Press.
Clutton-Brock, T.H., Albon, S.D., Gibson, R.M., and Guinness, F.E. (1979) 'The logical stag: adaptive aspects of fighting in red deer (*Cervus elaphus* L.), *Animal Behaviour* **27,** 211–25.
Clutton-Brock, T.H., Albon, S.D., and Guinness, F.E. (1981) 'Parental investment in male and female offspring in polygynous mammals', *Nature* **289**, 487–9.
Clutton-Brock, T.H., Albon, S.D., and Harvey, P.H. (1980) 'Antlers, body size and breeding group size in the Cervidae', *Nature* (Lond.) **285**, 565–7.
Clutton-Brock, T.H., Guinness, F.E., and Albon, S.D. (1982) *Red Deer: Behaviour and Ecology of Two Sexes.* Edinburgh University Press/Chicago University Press.
Clutton-Brock, T.H., and Harvey, P.H. (1978) 'Cooperation and Disruption'. In: *Readings in Sociobiology*, eds. Clutton-Brock, T.H., and Harvey, P.H., pp. 135–41. W.H. Freeman.
—— (1983) 'The functional significance of variation in body size among mammals'. *Special Publication of the American Society of Mammalogists* **7**, 632–63.
Clutton-Brock, T.H., Iason, G.R., and Guinness F.E. (1987) 'Sexual segregation and density-related changes in habitat use in male and female red deer', *J. Zool.* **211**, 275–89.
Clutton-Brock, T.H., Major, M., Albon, S.D., and Guinness, F.E. (1987) 'Early development and population dynamics in red deer. I. Density–dependent effects on juvenile survival', *J. Anim. Ecol.* **56,** 53–68.
Cooke, A.J., and Farrell, L. (1981) *The Ecology of the Chinese Water Deer* (Hydropotes inermis) *on Woodwalton Fen National Nature Reserve.* Nature Conservancy Council, Huntingdon.
Coupland, R.T. (1979) *Grassland Ecosystems of the World: Analysis of Grasslands and their Uses.* Cambridge University Press.
Cousins, M. (1987) An investigation of roe deer habitat usage and associated tree damage in a Scottish estate of mixed woodland type. BSc Hons thesis (Environmental Sciences), University of Southampton.
Corbet, G.B. (1978) *The Mammals of the Palearctic Region: A Taxonomic Review.* British Museum, London.
Corbet, G.B. and Southern, H.N. (1981) *The Handbook of British Mammals* (2nd edn). Blackwell, Oxford.
Cowan, I.McT. (1950) 'Some vital statistics of big game on overstocked mountain range', *Trans. N. Am, Wildl. Conf.* **15**, 581–8.
—— (1962) 'Hybridisation between the black-tailed and the white-tailed deer', *J. Mammal.* **43**, 539–41.
Cowan, I.McT., and Holloway, C.W. (1978) Introductory paper. In: IUCN, *Threatened Deer*, pp. 11–22.
Cowan, R.L., and Long, T.A. (1962) 'Studies on antler growth and nutrition of white-tailed deer'. Paper 107, Pennsylvania Cooperative Wildlife Research Unit, University Park, Pennsylvania.
Crawley, M.J. (1983) *Herbivory: The Dynamics of Animal-Plant Interactions.* Blackwell, Oxford.
Cumming, H.G. (1966) Behaviour and Dispersal in the Roe deer. PhD thesis, University of Aberdeen.
Dansie, O. (1973) *Muntjac.* British Deer Society.
Dapson, R.W., Ramsey, P.R., Smith M.K., and Urbston, D.F. (1979) 'Demographic differences in contiguous populations of white-tailed deer', *J. Wildl. Mgmt* **43**, 889–98.
Darwall, G. (1985) '250 years ago: a look back at deer natural history', *Deer* **6**, 260–1.
Darwall, G.H.D., and Clark, R.G. (1986) 'On the growth and form of deer antlers', *Deer* **6**, 341–4
Dasmann, R.F., and Taber, R.D. (1956) 'Behaviour of Columbian black-tailed deer with reference to population ecology', *J. Mammal.* **37**, 143–64.
David, Père (1987) *Deer to China.* Thouless & Loudon.

de Bie, S. (1988) Ruminal fermentation rates and feeding strategies in ruminants. (*In press*)
de Bie, S., Joenje, W., and van Wieren, S.E. (1988) *Begrazing door Vertebraten*. Pudoc, Netherlands.
Delap, P. (1978) *Roe Deer*. British Deer Society.
De Nahlik, A.J. (1959) *Wild Deer*. Faber and Faber, London.
Demment, M.W., and Van Soest, P.J. (1985) 'A nutritional explanation for body size patterns of ruminant and non-ruminant herbivores, *American Naturalist* **125**, 641–72.
Dinesman, L.G. (1967) 'Influence of vertebrates on primary production of terrestrial communities'. In: *Secondary Productivity in Terrestrial Ecosystems*, ed. Petrusewicz, K., Vol. 1, 261–6.
Downing, R.L., and McGinnes, B.S. (1969) 'Capturing and marking white-tailed deer fawns', *J. Wildl. Mgmt* **33**, 711–14.
Dratch, P., and Gyllensten, U. (1985) 'Genetic differentiation of red deer and North American elk (wapiti)'. In: *Biology of Deer Production*, eds. Fennessy, P.F., and Drew, K.R., pp. 37–40. Royal Society of New Zealand.
Drew, K.R. (1985) 'Meat production from farmed deer. In: *Biology of Deer Production*, eds. Fennessy, P.F., and Drew, K.R., pp. 285–90. Royal Society of New Zealand.
Drozdz, A. (1979) 'Seasonal intake and digestibility of natural foods by roe deer', *Acta Theriologica* **24**, 137–70.
Drozdz, A., Weiner, J., Gebczynska, Z., and Krasinska, M. (1975) 'Some bioenergetic parameters of wild ruminants', *Polish Ecological Studies* **1**, 85–101.
Dunkeson, R.L., and Murphy, D.A. (1953) 'Missouri's deer herd: reproduction and checking station data'. Missouri Conservation Commission, Jefferson City.
Eadie, R. (1961) 'Control of wildlife damage in orchards'. Bull. 1055, New York State Agric. Ext. Station, Ithaca.
Eberhardt, L. (1969) 'Population analysis'. In: *Wildlife Management Techniques*, ed. Giles, R.H., pp. 457–95. W.H. Freeman.
Egorov, O.V. (1965) *Wild Ungulates of Yakutia*. Nauka. Moscow. (Translated from Russian by the Israel Program for Scientific Translations, Jerusalem, 1967.)
Eisenberg, J.F., and Lockhart, M. (1972) 'An ecological reconnaissance of Wilpattu National Park, Ceylon, *Smithsonian Contributions to Zoology* **101**, 1–118.
Ellenberg, H. (1978) 'The population ecology of roe deer, *Capreolus capreolus* (Cervidae) in Central Europe', *Spixiana* Suppl. **0**(2), 5–211.
Ellerman, J.R., and Morrison-Scott, T.C.S. (1951) *Check List of Palearctic and Indian Mammals 1758–1946*. British Museum (Nat. Hist.), London.
Ellison, L. (1960) 'Influence of grazing on plant succession of rangelands', *Botanical Review* **26**, 1–78.
Eltringham, S.K. (1984) *Wildlife Resources and Economic Development*. Wiley International.
Estes, R.D. (1974) 'Social organisation of the African Bovidae'. In: *The Behaviour of Ungulates and its Relationship to Management*, eds. Geist, V., and Walther, F. IUCN, Switzerland.
Fennessy, P.F. (1981) 'Nutrition of red deer'. Proceedings of a Deer Seminar for Veterinarians, Queenstown, N.Z. Vet. Assoc., pp. 8–16.
Fennessy, P.F., and Drew, K.R. (eds.) (1985) *Biology of Deer Production*. Royal Society of New Zealand Bulletin **22**, 1–482.
Fennessy, P.F., and Suttie, J.M. (1985) 'Antler growth: nutritional and endocrine factors'. In: *Biology of Deer Production*, eds. Fennessy, P.F., and Drew, K.R., pp. 239–50. Royal Society of New Zealand.
Fischer, H. (1975) [Sugar licks prove successful]. *Die Pirsch – Der Deutsche Jäger*, Vol. 3.
Flerov, C.C. (1930) 'On the classification and geographical distribution of the genus *Moschus* (Mammalia: Cervidae)', *Ann. Mus. Zool. Acad. Sci. USSR* **31**, 1–20.
—— (1952) *Fauna of the USSR. I. Mammals: Musk Deer and Deer*. USSR Academy of Sciences, Moscow.
Fletcher, T.J. (1975) The environmental and hormonal control of reproduction in male and female red deer (*Cervus elaphus*). PhD thesis, University of Cambridge.
—— (1978) 'The induction of male sexual behaviour in red deer (*Cervus elaphus*) by the administration of testosterone to hinds and estradiol – 17 B to stags', *Horm. Behav.* **11**, 74–88.
Flook, D.R. (1970) *A study of sex differential in the survival of wapiti*. Canadian Wildlife Reports No. 11. Department of Indian Affairs and Northern Development, Ottawa.
Flower, W.H. (1875) 'On the structure and affinities of the musk deer (*Moschus moschiferus* Linn.)', *Proc. Zool. Soc. Lond.* 159–90.

Flyger, V.F., and Thoerig, T. (1965) 'Crop damage caused by Maryland deer', *Proc. Ann. Conf. Southeast Assoc. Game & Fish. Comm.* **16**, 45–52.
Franklin, W.L., and Lieb, J.W. (1979) 'The social organisation of a sedentary population of North American elk: a model for understanding other populations'. In: *North American Elk: Ecology, Behaviour and Management*, eds. Boyce, M.S., and Hayden-Wing, L.D., pp. 185–98. University of Wyoming Press.
Franklin, W.L., Mossman, A.S., and Dole, M. (1975) 'Social organisation and home range of Roosevelt elk', *J. Mammal.* **56**, 102–18.
Franzmann, A.W. (1978) 'Moose' in: Big Game of North America, eds Schmidt, J.L. and Gilbert, D.L., pp. 67–82. Stackpole Books, Pennsylvania.
Fraser Stewart, J.W. (1981) 'A study of the wild rusa deer (*Cervus timorensis*) in the Tonda Wildlife Management Area, Papua New Guinea'. PNG/78/040 Field document No. 1. Wildlife Division, Dept Lands and Environment. FAO, Port Moresby.
Gallini, S. (1984) 'Ecological aspects of the co-exploitation of white-tailed deer and cattle', *Acta Zool. Fennica* **172**, 251–4.
Gasaway, W.C., and Coady, J.W., (1974) 'Review of energy requirements and rumen fermentation in moose and other ruminants', *Naturaliste Canadien* **101**, 227–62.
Gauthier, D., and Barrette, C. (1986) 'Suckling and weaning in captive white-tailed and fallow deer', *Behaviour* **94**, 128–49.
Geiger, C., and Kramer, A. (1974) 'Rangordnung von Rehwild (*Capreolus capreolus* L.) am der Winterfutterung in einem schweizerischen Jagdrevier', *Z. Jagdwiss.* **20**, 53–6.
Geist, V. (1971) *Mountain Sheep: A Study in Behaviour and Evolution.* University of Chicago Press.
—— (1974) 'On the relationship of social evolution and ecology in ungulates', *Amer. Zool.* **14**, 205–20.
Geist, V., and Bromley, P.T. (1978) 'Why deer shed antlers', *Z. Säugetierkunde* **43**, 223–31.
Geist, V., and Petocz, R.G. (1977) 'Bighorn sheep in winter: do rams maximise reproductive fitness by spatial and habitat segregation from ewes', *Canadian J. Zool.* **55**, 1802–10.
Gent, A.H. (1983) Range use and activity patterns in roe deer (*Capreolus capreolus*): a study of radio-tagged animals in Chedington Wood, Dorset. BSc Hons thesis (Biology), University of Southampton.
Gordon, C. (1986) Time-budgetting, suckling behaviour and changes in group size in a captive herd of fallow deer (*Dama dama*). MSc thesis, University of Durham.
Gordon, I.J., and Illius, A.W. (1988) 'Incisor arcade structure and diet selection in herbivores', *Functional Ecology*, in press.
—— (1983) *Deer Antlers: Regeneration, Function, and Evolution.* Academic Press.
Goss, R.J. (1985) 'Tissue differentiation in regenerating antlers'. In: *Biology of Deer Production*, eds. Fennessy, P.F., and Drew, K.R., pp. 229–38. Royal Society of New Zealand.
Gould, S.J. (1973) 'Positive allometry of antlers in the "Irish Elk", *Megaloceros giganteus'*, *Nature* (Lond.) **244**, 375–6.
—— (1977). *Ever Since Darwin: Reflections in Natural History.* Pelican Books.
Grant, S.A., and Hunter, R.F. (1966) 'The effects of frequency and season of clipping on the morphology, productivity and chemical composition of *Calluna vulgaris'*, *New Phytologist* **65**, 125–33.
Gray, A.J., and Scott, R. (1977) '*Puccinellia maritima* (Huds) Par. 1', *J. Ecol.* **65**, 699–716.
Gray, A.P. (1971) *Mammalian Hybrids.* Commonwealth Agriculture Bureau, Edinburgh.
Gray, J.E. (1821) 'On the natural arrangement of vertebrate mammals', *London Med. Reposit.* **15**, 296–310.
Green, M.J.B. (1985) Aspects of the ecology of the Himalayan musk deer. PhD thesis, University of Cambridge.
—— (1986) 'The distribution, status and conservation of the Himalayan musk deer (*Moschus chrysogaster*)', *Biol. Conserv.* **35**, 347–75.
—— (1987a) 'Ecological separation in Himalayan ungulates', *J. Zool.* (Lond.) (B) **1**, 693–719.
—— (1987b) 'Exploiting the musk deer for its musk', *Traffic Bulletin* (IUCN) **8**, 59–66.
—— (1987c) 'Diet composition and quality in Himalayan musk deer based on fecal analysis', *J. Wildl. Mgmt* **51**, 880–92.
—— (1982) 'The species of muntjac (genus *Muntiacus*) in Borneo: unrecognised sympatry in tropical deer', *Zoologische Mededelingen* (Leiden) **56**, 203–16.
Groves, C.P., and Grubb, P. (1987) 'Relationships of living Cervidae'. In: *Biology and Management of the Cervidae*, ed. Wemmer, C. Smithsonian, Washington.
Guinness, F.E., Clutton-Brock, T.H., and Albon, S.D. (1978) 'Factors affecting calf mortality in red deer', *J. Anim. Ecol.* **47**, 817–32.

Guinness, F.E., and Fletcher, J. (1971) 'First ever recorded incidence of twins born to a red deer hind in Britain', *Deer* **2**, 680–3.
Gulland, J.A. (1962) 'The application of mathematical models to fish populations'. In: *The Exploitation of Natural Animal Populations*, eds. Le Cren, E.D., and Holdgate, M.W., pp. 204–18. Wiley, New York.
Hall, W. (1987) 'Collection news', *Marwell Zoo Newsletter* **54**, 2–3.
Halls, L.K. (ed.) (1984) *White-Tailed Deer: Ecology and Management.* Wildlife Institute of America, Stackpole Books, Pennsylvania.
Hamilton, E. (1871) 'Remarks on the prolific nature of *Hydropotes inermis*', *Proc. Zool. Soc. Lond.* (1871), 258.
Hamilton, W.J., and Blaxter, K.L. (1980) 'Reproduction in farmed red deer. I. Hind and stag fertility', *J. Agric. Sci. Camb.* **95**, 261–73.
Hanley, T.A. (1984) 'Habitat patches and their selection by wapiti and black-tailed deer in a coastal montane coniferous forest', *J. Appl. Ecol.* **21**, 423–36.
Hanley, T.A., and Hanley, K.A. (1982) 'Food resource partitioning by sympatric ungulates on Great Basin rangeland', *J. Range Mgmt* **35**, 152–8.
Hansen, R.M., Clark, R.C., and Lawhorn, W. (1977) 'Food of wild horses, deer and cattle in the Douglas Mountain area, Colorado', *J. Range Mgmt* **30**, 116–18.
Hansen, R.M., and Reid, L.D. (1975) 'Diet overlap of deer, elk and cattle in southern Colorado', *J. Range Mgmt* **28**, 43–7.
Harder, J.D. (1985) 'Exploitation of a high-density herd of white-tailed deer'. In: *Biology of Deer Production*, eds. Fennessy, P.F., and Drew, K.R., pp. 45–8. Royal Society of New Zealand.
Harrington, R. (1982) 'The hybridization of red deer (*Cervus elaphus* L. 1758) and Japanese sika deer (*Cervus nippon* Temminck 1838)', *Trans. Int. Congr. Game Biol.* **14**, 559–71.
—— (1985) 'Evolution and distribution of the Cervidae'. In: *Biology of Deer Production*, eds. Fennessy, P.F., and Drew, K.R., pp. 3–11. Royal Society of New Zealand.
Harris, S., and Forde, P. (1986) 'The annual diet of muntjac (*Muntiacus reevesi*) in the King's Forest, Suffolk', *B.E.S. Bull.* **17**, 19–22.
Harrison, R.J., and Hyett, A.R. (1954) 'The development and growth of the placentomes in the fallow deer (*Dama dama* L,)', *J. Anat. Lond.* **88**, 338–55.
Hawkins, D. (1987) The parasitic interrelationships of sheep and deer on Knebworth House Estate, near Stevenage. BSc. Hons thesis (Biology), University of Southampton.
Hayne, D.H. (1984) 'Population dynamics and analysis'. In: *White-tailed Deer, Ecology and Management*, ed. Halls, L.K., pp. 203–10. Stackpole Books, Pennsylvania.
Heady, H.F. (1964) 'Palatibility of herbage and animal preference', *J. Range Mgmt* **17**, 76–82.
Henry, B.A.M. (1978) 'Diet of roe deer in an English conifer forest', *J. Wild. Mgmt* **42**, 937–9.
—— (1981) 'Distribution patterns of roe deer (*Capreolus capreolus*) related to the availability of food and cover', *J. Zool.* **194**, 271–5.
Heptner, W.A., Nasimovitsch, A.A., and Bannikov, A.G. (1961) *Mammals of the Soviet Union.* Fischer-Verlag, Jena.
Hesselton, W.T., and Jackson, L.W. (1971) 'Some reproductive anomalies in female white-tailed deer from New York', *New York Fish and Game J.* **18**, 42–51.
—— (1977) 'Reproductive rates of white-tailed deer in New York state', *New York Fish and Game J.* **21**, 135–52.
Hilborn, R., and Sinclair, A.R.E. (1979) 'A simulation of the wildebeest population, other ungulates, and their predators'. In: *Serengeti: Dynamics of an Ecosystem*, eds. Sinclair, A.R.E., and Norton-Griffiths M., pp. 287–309. Chicago University Press.
Hinge, M.D.C. (1986) Ecology of red and roe deer in a mixed-age conifer plantation: comparative studies on habitat selection, ranging behaviour and feeding strategies. PhD thesis, University of Aberdeen.
Hirth, D.H. (1973) Social behaviour of white-tailed deer in relation to habitat. PhD thesis, University of Michigan, Ann Arbor.
—— (1977). 'Social behaviour of white-tailed deer in relation to habitat', *Wildl. Monographs* **53**, 55 pp.
Hofmann, R.R. (1973) *The Ruminant Stomach.* East African Literature Bureau, Nairobi.
—— (1976) 'Zur adaptiven Differenzierung der Wiederkauer: Untersuchungsergebnisse auf der basis der vergleichenden funktionellen Anatomie des verdauungstrakts', *Prakt. Tierarzt.* **57**, 351–8.
—— (1985) 'Digestive physiology of the deer — their morphophysiological specialisation and adaptation'. In: *Biology of Deer Production*, eds. Fennessy, P.F., and Drew, K.R, pp. 393–407. Royal Society of New Zealand.

Hofmann, R.R., and Schnorr, R.R.B. (1982) *Funktionelle Morphologie des Wiederkauer-Magens (Schleimhaut und Versorgungsbahnen)* Ferd. Enke-Verlag, Stuttgart.

Hofmann, R.R., and Stewart, D.R.M. (1972) 'Grazer or browser? a classification based on the stomach structure and feeding habits of East African ruminants', *Mammalia* **36**, 226–40.

Holisova, V., Kozena, I., and Obrtel, R. (1984) 'The summer diet of field roe bucks (*Capreolus capreolus*) in Southern Moravia', *Folia Zoologica* (Brno) **33**, 193–208.

Holloway, J.T. (1950) 'Deer and the forests of Western Southland', *N. Z. J. Forestry* **6**, 123–37.

Holmes, F. (1974) *Following the Roe*. Bartholomew and Sons, Edinburgh.

Holt, M.E. (1976) 'An investigation of the internal parasite burden of the deer and associated animals at Knebworth Park, Herts, *Deer* **4**, 40–1.

Holter, J.B., Urban, W.E., and Hayes, H.H. (1977) 'Nutrition of northern white-tailed deer throughout the year', *J. Anim. Science* **45**, 365–76.

Hope Simpson, J.F. (1940) 'Studies of the vegetation of the English Chalk. VI. Late stages in succession leading to chalk grassland', *J. Ecol.* **28**, 386–402.

Hoppe, P.P., Qvortrup, S.A., and Woodford, M.H. (1977) 'Rumen fermentation and food selection in East African sheep, goats, Thomson's gazelle, Grant's gazelle and impala', *J. Agric. Sci. Cambr.* **89**, 129–35.

Horn, E.E. (1941) 'Some coyote-wildlife relationships', *Trans. N. Am. Wildl. Conf.* **6**, 283–7.

Hornocker, M.G. (1970) 'An analysis of mountain lion predation upon mule deer and elk in the Idaho primitive area', *Wildlife Monographs* **21**, 1–39.

Horwood, M.T., and Masters, E.H. (1970) 'Sika Deer (*Cervus nippon*) (with particular reference to the Poole Basin)'. British Deer Society. 30 pp.

Hosey, G.R. (1974) The food and feeding ecology of roe deer. PhD thesis, University of Manchester.

—— (1981) 'Annual foods of the Roe deer (*Capreolus capreolus*) in the south of England', *J. Zool.* **194**, 276–9.

Houston, D.B. (1982) *The Northern Yellowstone Elk: Ecology and Management*. Macmillan, New York.

Hsu, C.M., Na, C., Kan, W.S., and Yiu, C.S. (1979) 'Studies of Pharmaceutical zoology on crude drugs of Cervidae (deer) origin in Taiwan', *China Medical College Annual Bulletin* **10**, 405–76.

Illius, A.W., and Gordon, I.J. (1987) 'The allometry of food intake in grazing ruminants', *J. Anim, Ecol.* **56**, 989–99.

IUCN (1978) *Threatened Deer*. Morges, Switzerland.

Jackes, A.D. (1973) The use of wintering grounds by red deer in Ross-shire, Scotland. PhM dissertation, University of Edinburgh.

Jackson, J.E. (1974) The feeding ecology of fallow deer in the New Forest, Hampshire. PhD thesis, University of Southampton.

—— (1977a) 'The annual diet of the fallow deer (*Dama dama*) in the New Forest, Hampshire, as determined by rumen content analysis', *J. Zool. Lond.* **181**, 465–73.

—— (1977b) 'When do fallow deer feed?', *Deer* **4**, 215–18.

—— (1977c) 'The duration of lactation in New Forest fallow deer', *J. Zool. Lond.* **183**, 542–3.

—— (1978) 'The Argentinian pampas deer or venado (*Ozotoceros bezoarticus celer*)'. In IUCN, *Threatened Deer*, pp. 33–45.

—— (1980) 'The annual diet of the roe deer (*Capreolus capreolus*) in the New Forest Hampshire, as determined by rumen content analysis', *J. Zool. Lond.* **192,** 71–83.

Jackson, J., Landa, P., and Langguth, A. (1980) 'Pampas deer in Uruguay', *Oryx* **15**, 267–72.

Jackson, R.M., White, M., and Knowlton, F.F. (1972) 'Activity patterns of young white-tailed deer fawns in south Texas', *Ecology* **53,** 262–70.

Jaczewski, Z. (1976) 'The induction of antler growth in female red deer', *Bull. Acad, Polon. Sci.* **24**, 61–5.

Jarman, P.J. (1974) 'The social organisation of antelope in relation to their ecology', *Behaviour* **48**, 215–66.

Jarman, P.J., and Sinclair, A.R.E. (1979) 'Feeding strategy and the pattern of resource-partitioning in ungulates'. In: *Serengeti: Dynamics of an Ecosystem*, eds. Sinclair A.R.E., and Norton-Griffiths, M., pp. 130–63. Chicago University Press.

Jenkins, K.J., and Wright, R.G. (1988) 'Resource partitioning and competition among cervids in the northern Rocky Mountains', *J. Appl. Ecol.*, in press.

Johnson, T.H. (1984) Habitat and social organisation of roe deer (*Capreolus capreolus*). PhD thesis, University of Southampton.

Joubert, D.M. (1963) 'Puberty in farm animals', *Animal Breeding Abstracts* **31**, 295–306.

Julander, O., Robinette, W.L., and Jones, D.A. (1961) 'Relation of summer range condition to mule deer herd productivity', *J. Wildl. Mgmt* **25**, 54–60.

Kaluzinski, J. (1982a) 'Dynamics and structure of a field roe deer population', *Acta Theriologica* **27**, 385–408.

—— (1982b) 'Composition of the food of roe deer living in fields and the effects of their feeding on plant production', *Acta Theriologica* **27**, 457–70.

Kammermeyer (1975) Movement – ecology of White-tailed deer in relation to a refuge and hunted area. MS thesis, University of Georgia.

Katsma, D.E., and Rusch, D.H. (1980) 'Effects of simulated deer-browsing on branches of apple-trees', *J. Wildl. Mgmt* **44**, 603–12.

Kay, R.N.B. (1979) 'Seasonal changes of appetite in deer and sheep.' Agricultural Research Council Research Review, 1979.

—— (1985) 'Body size, patterns of growth and efficiency of production in red deer'. In: *Biology of Deer Production*, eds. Fennessy, P.F., and Drew, K.R., pp. 411–21. Royal Society of New Zealand.

Kay, R.N.B., and Staines, B.W. (1981) 'The nutrition of the red deer (*Cervus elaphus*)', *Nutrition Abstracts and Reviews, Series B*, **51**, 601–21.

Kean, R.I. (1959) 'Ecology of the larger wildlife mammals of New Zealand', *N. Z. Sci. Reviews* **17**, 35–7.

Kelsall, J.P. (1968) *The Migratory Barren-ground Caribou of Canada* Canadian Wildlife Service.

Kiddie, D.G. (1962) *The sika deer* (Cervus nippon) *in New Zealand.* N. Z. Forest Service, Information Series. No. 44.

—— (1987) 'Fighting behaviour of the extinct Irish elk', *Modern Geology* **11**, 1–28.

Kitchener, A.C. (1988) 'The evolution and mechanical design of horns and antlers'. In: *Biomechanics in Evolution* ed. Rayner, Chapter 12. In press.

Klein, D.R. (1968) 'The introduction, increase and crash of reindeer on St. Matthew Island', *J. Wildl. Mgmt* **32**, 350–67.

Klein, D.R., and Olson, S.T. (1960) 'Natural mortality patterns of deer in southeast Alaska', *J. Wildl. Mgmt* **24**, 80–8.

Kong, Y.C., and But, P.P.H. (1985) 'Deer — the ultimate medicinal animal: (Antler and deer parts in medicine)'. In: *Biology of Deer Production*, eds. Fennessy, P.F., and Drew, K.R., pp. 311–24. Royal Society of New Zealand.

Krefting, L.W., Stenlund, M.H., and Seemel, R.K. (1966) 'Effect of simulated and natural browsing on mountain maple', *J. Wildl. Mgmt* **30**, 481–8.

Krishnan, M. (1959) *The Mudamalai Wildlife Sanctuary*. Madras State Forest Dept.

Kruuk, H. (1972) *The Spotted Hyena – a Study of Predation and Social Behaviour*. Chicago University Press.

Kruuk, H., and Turner, M. (1967) 'Comparative notes on predation by lion, leopard, cheetah and wild dog in the Serengeti area, East Africa, *Mammalia* **31**, 1–27.

Kurt, F. (1978) 'Socio–ecological organisation and aspect of management in South Asian deer'. In: *Threatened Deer*, pp. 219–39. IUCN, Switzerland.

Lagory, K.E. (1986) 'Habitat, group size and behaviour of white-tailed deer', *Behaviour* **98**, 168–79.

Langenau, E.E., and Lerg, J.M. (1976) 'The effects of winter nutritional stress on maternal and neonatal behaviour in penned white-tailed deer', *Appl. Anim. Ethol.* **2**, 207–23.

Leader-Williams, N. (1980) 'Population dynamics and mortality of reindeer introduced into South Georgia', *J. Wildl. Mgmt* **44**, 640–57.

—— (1988) *Reindeer on South Georgia: The Ecology of an Introduced Population.* Cambridge University Press.

Lincoln, G. (1971) 'Puberty in a seasonally breeding male, the red deer stag, (*Cervus elaphus*)', *J. Reprod. Fert.* **25**, 41–54.

Lincoln, G.A. (1984) 'Antlers and their regeneration – a study using hummels, hinds and haviers', *Proc. Roy. Soc. Edinburgh* **82**B, 243–59.

—— (1985). 'Seasonal breeding in deer'. In: *Biology of Deer Production*. eds. Fennessy, P.F., and Drew, K.R., pp. 165–80. Royal Society of New Zealand.

Lincoln, G.A., and Bubenik, G.A. (1985) 'Antler physiology'. In: *Biology of Deer Production*, eds. Fennessy, P.F., and Drew, K.R., pp. 474–6. Royal Society of New Zealand.

Lincoln, G., and Fletcher, J. (1969) 'History of a hummel. I', *Deer* **1**, 327.

—— (1970) 'History of a hummel. II', *Deer* **2**, 630.

—— (1973) 'History of a hummel. III. Sons with antlers', *Deer* **3**, 26.

—— (1976) 'History of a hummel. IV. The hummel dies', *Deer* **3**, 552–5.

—— (1977) 'History of a hummel. V. Offspring from father/daughter matings', *Deer* **4**, 86–7.

—— (1978) 'History of a hummel. VI. Half a hummel', *Deer* **4**, 274–5.

—— (1984) 'History of a hummel. VII. Nature vs nurture', *Deer* **6**, 127–31.

Lincoln, G.A., Guinness, F., and Short, R.V. (1972) 'The way in which testosterone controls the social and sexual behaviour of the red deer stag (*Cervus elaphus*)', *Horm. Behav.* **3**, 375–96.
Loudon, A.S.I. (1979) Social behaviour and habitat in roe deer (*Capreolus capreolus*). PhD thesis, Edinburgh University.
—— (1987) 'The influence of forest habitat structure on growth, body size and reproduction in roe deer (*Capreolus capreolus* L.)'. In: *Biology and Management of the Cervidae*, ed. Wemmer, C.M., pp. 559–69. Smithsonian, Washington.
Loudon, A.S.I., Darroch, A,D., and Milne, J.A. (1984) 'The lactation performance of red deer on hill and improved species pasture', *J. Agric. Sci. Camb.* **102**, 149–58.
Loudon, A.S.I., and Milne, J.A. (1985) 'Nutrition and growth of young red deer'. In: *Biology of Deer Production*, eds. Fennessy, P.F., and Drew, K.R., pp. 423–7. Royal Society of New Zealand.
Lowe, V.P.W. (1966) 'Observations on the dispersal of red deer on Rhum'. In: *Play, Exploration and Territory in Mammals*, eds. Jewell, P.A., and Loizos, C., pp. 211–28. Academic Press.
—— (1969) 'Population dynamics of the red deer (*Cervus elaphus* L.) on Rhum', *J. Anim. Ecol.* **38**, 425–57.
Lowe, V.P.W., and Gardiner, A.S. (1975) 'Hybridisation between red deer (*Cervus elaphus*) and sika deer (*Cervus nippon*) with particular reference to stocks in N.W. England', *J. Zool. Lond.* **177**, 553–66.
Lydekker, R. (1898) *The Deer of All Lands*. Rowland Ward, London.
Mann, J.C.E. (1983) The social organisation of ecology of the Japanese sika deer (*Cervus nippon*) in southern England. PhD thesis, University of Southampton.
Mansell, W.D. (1974) 'Productivity of white-tailed deer on the Bruce Peninsula, Ontario', *J. Wildl. Mgmt* **38**, 808–14.
Marchinton, R.L. (1964) Activity cycles and mobility of central Florida deer based on telemetric and observational data. MS thesis, University of Florida.
Marchinton, R.L., and Hirth, D.H. (1984) 'Behaviour'. In: *White-tailed deer: Ecology and Management*, ed. Halls, L.K., pp. 129–68. Stackpole Books, Pennsylvania.
Mark, A.F., and Baylis, G.T.S. (1975) 'Impact of deer on Secretary Island, Fiordland, New Zealand', *Proc. N.Z. Ecol. Soc.* **22**, 19–24.
Martin, C.. (1978) 'Status and ecology of the barasingha (*Cervus duvauceli branderi*) in Kanha National Park, India', *J. Bombay Nat. Hist. Soc.* **74**, 61–132.
Matschke, G.H., de Calesta, D.S., and Harder, J.D. (1984) 'Crop damage and control'. In: *White-tailed Deer: Ecology and Management*, ed. Halls, L.K., ch. 38. Stackpole Books, Pennsylvania.
Mautz, W.W. (1978) 'Nutrition and carrying capacity'. In: *Big Game of North America: Ecology and Management*, eds. Schmidt, J.L., and Gilbert, D.L., pp. 321–48. Stackpole Books, Pennsylvania.
MacArthur, R.H., and Levins, R. (1967) 'The limiting similarity, convergence and divergence of coexisting species', *Amer. Nat.* **101**, 377–85.
McCullough, D.R. (1979) *The George Reserve Deer Herd: Population Ecology of a K-selected Species*. University of Michigan Press, Ann Arbor.
—— (1982) 'Population growth rate of the George Reserve deer herd', *J. Wildl. Mgmt* **46**, 1079–82.
McDiarmid, A. (ed.) (1969) 'Diseases in free-living wild animals', *Symp. Zool. Soc. Lond.* **24**.
McEwan, E.K. (1968) 'Growth and development of barren ground caribou. II. Postnatal growth rates', *Can. J. Zool.* **46,** 1023–9.
McEwan, E.K., and Whitehead, P.E. (1970) 'Seasonal changes in the energy and nitrogen intake in reindeer and caribou', *Can. J. Zool.* **48**, 905–13.
McEwen, L.S., French, C.E., Magruder, N.D., Swift, R.W., and Ingram, R.H. (1959) 'Nutrient requirements of the white-tailed deer', *Trans. 22nd N. Am. Wildlife Conference*, 119–32.
McKelvey, P.J. (1959) 'Animal damage in North Island protection forests', *N.Z. Science Review* **17**, 28–34.
McMichael, T.J. (1970) 'Rate of predation on deer fawn mortality'. In: *Wildlife Research in Arizona 1969–70*, pp. 77–83. Arizona Fish and Game Dept, Phoenix.
MacNally, L. (1982) 'The observed birth of twins to a red deer hind', *Deer* **5**, 451–3.
McNaughton, S.J. (1979) 'Grazing as an optimisation process: grass-ungulate relationships in the Serengeti', *Amer. Nat.* **113**, 691–703.
—— (1979) 'Grassland-herbivore dynamics'. In: *Serengeti: Dynamics of an Ecosystem*, eds. Sinclair, A.R.E., and Norton-Griffiths, M., pp. 46–81. University of Chicago Press.

Mech, L.D. (1966) *The Wolves of Isle Royale.* Fauna of the National Parks No. 7. US Government Printing Office, Washington.
—— (1970) *The Wolf.* Natural History Press/Doubleday, New York.
Mech, L.D., and Karns, P.D. (1977) *Role of the Wolf in a Deer Decline in the Superior National Forest.* USDA Forest Service, Research Paper NC–148, 1–23.
Messier, F., and Crête, M. (1985) 'Moose-wolf dynamics and the natural regulation of moose populations', *Oecologia* **65**, 503–12.
Middleton, A.D. (1937) 'Whipsnade ecological survey: 1936–7', *Proc Zool. Soc. Lond.* **107**, 471–81.
Miquelle, D.G. (1983) 'Browse regrowth and consumption following summer defoliation by moose', *J. Wildl. Mgmt* **47**, 17–24.
Mishra, H.R. (1982) The ecology and behaviour of chital (*Axis axis*) in the Royal Chitawan National Park, Nepal. PhD thesis, University of Edinburgh.
Mishra, R.H., Arora, R.B., and Seth, S.D.S. (1962) 'Anti-inflammatory activity of musk', *J. Pharm. Pharmac.* **14**, 830–1.
Mitchell, B., and Brown, D. (1974) 'The effects of age and body size on fertility in female red deer (*Cervus elaphus* L.)', *Proc. Int. Congr. Game, Biol.* **11**, 89–98.
Mitchell, B., Staines, B.W., and Welch, D. (1977) *Ecology of Red Deer: A Research Review Relevant to their Management.* Institute of Terrestrial Ecology, Cambridge.
Montgomery, G.G. (1963) 'Nocturnal movements and acitivity rhythms of white-tailed deer', *J. Wildl. Mgmt* **27**, 422–7.
Morton, G.H., and Chaetum, E.L. (1946) 'Regional differences in breeding potential of white-tailed deer of New York', *J. Wildl. Mgmt* **10**, 242–8.
Muir, P.D., Sykes, A.R., and Barrell, G.K. (1985) 'Mineralisation during antler growth in red deer'. In: *Biology of Deer Production*, eds. Fennessy, P.F., and Drew, K.R. pp. 251–4. Royal Society of New Zealand.
Mukerji, B. (1953) *Indian Pharmaceutical Codex,* Vol. 1. Council of Scientific and Industrial Research, New Delhi.
Mukhopadhyay, A., Seth, S.D.S., and Bagchi, N. (1973) 'Cardiac and CNS actions of musk', *Indian. J. Pharm.* **35**, 169–70.
Murphy, D.A., and Coates, J.A. (1966) 'Effects of dietary protein on deer', *Trans. N. Am. Wildl. and Natur. Resource Conf.* **31**, 129–39.
Newton, P.N. (1984) The ecology and social organisation of Hanuman langurs (*Presbytis entellus* Dufresne 1797) in Kanha Tiger Reserve, Central India Highlands. D.Phil thesis, University of Oxford.
Nicholson, I.A., Paterson, I.S., and Currie, A. (1970) 'A study of vegetational dynamics: selection by sheep and cattle in *Nardus* pasture'. In: *Animal Populations in Relation to their Food Resources*, ed. Watson, A. BES Symposium **10**, 129–43.
Obrtel, R., and Holisova, V. (1983) 'Assessment of the damage done to a crop of maize (*Zea mays*) by roe deer (*Capreolus capreolus*)', *Folia Zoologica* (Brno) **32**, 109–18.
Obrtel, R., Holisova, V., and Kozena, I. (1984) 'Deer damage to sugar beet leaves', *Folia Zoologica (Brno)* **33**, 99–108.
Ozoga, J.J., Verme, L.J., and Bienz, C.S. (1982) 'Parturition behaviour and territoriality in white-tailed deer: impact on neonatal mortality', *J. Wildl. Mgmt* **46**, 1–11.
Peterson, R.L. (1955) *North American Moose.* University of Toronto Press.
Peterson, R.O. (1977) *Wolf Ecology and Prey Relationships on Isle Royale.* Scientific Monograph Series No. 11, US National Park Service.
Pianka, E.R. (1973) 'The structure of lizard communities', *Ann. Rev. Ecol. Syst.* **4**, 53–74.
Pielowski, Z. (1969) 'Die Wiedereinburgerung des Elches (*Alces alces* L.) im Kampinos-National park in Polen', *Z. Jägdwissenschaft* **15**, 6–17.
Pimlott, D.H. (1967) 'Wolf predation and ungulate populations', *Amer. Zool.* **7**, 267–78.
Pitcher, T. (1979) 'He who hesitates, lives. Is stotting anti-ambush behaviour?', *Amer. Nat.* **113**, 453–6.
Poutsma, J. (1987) 'Social preferences by roe bucks when joining family groups in winter', *Deer* **7**, 73–6.
Povilitis, A. (1978) 'The Chilean Huemul Project'. In: *Threatened Deer*, pp. 109–. IUCN, Switzerland.
Powerscourt, Viscount (1884) 'On the acclimatization of the Japanese deer at Powerscourt', *Proc. Zool. Soc. Lond.* 207–9.
Prater, S. (1934) *The Wild Animals of the Indian Empire.* Madras.
—— (1945) 'Breeding habits of swamp deer (*Rucervus duvauceli*) in Assam', *J. Bombay Nat. Hist. Soc.* **45**, 415–16.
Prior, R. (1968) *The Roe Deer of Cranborne Chase.* Oxford University Press.

Putman, R.J. (1981) 'Social systems of deer: a speculative review', *Deer* **5**, 186–8.
—— (1985) 'Humane control of land mammals and birds: alternatives to direct population control'. In: *Humane Control of Land Mammals and Birds*, pp. 124–31. Universities Federation for Animal Welfare.
—— (1986a) *Grazing in Temperate Ecosystems: Large Herbivores and the Ecology of the New Forest.* Croom Helm, Beckenham.
—— (1986b) 'Competition and coexistence in a multi-species grazing system', *Acta Theriologica* **31**, 271–91.
—— (1986c) 'Foraging by roe deer in agricultural areas and impact on arable crops', *J. Appl. Ecol.* **23**, 91–100.
Putman, R.J., and Wratten, S.D. (1984) *Principles of Ecology.* Croom Helm, Beckenham.
Radwan, M.A., and Crouch, G.L. (1974) 'Plant characteristics related to feeding preference by black-tailed deer', *J. Wildl. Mgmt* **38**, 32–14.
Raedeke, K.J., and Taber, R.D. (1985) 'Do cougars reduce hunter harvest of black-tailed deer in Washington State?' In: *Biology of Deer Production*, ed. Fennessy, P.F., and Drew, K.R., pp. 49–50. Royal Society of New Zealand.
Ranjitsingh, M.K. (1978) 'The Manipur brow-antlered deer (*Cervus eldi eldi*) – a case history'. In: *Threatened Deer*, pp. 26–32. IUCN, Switzerland.
Ransom, A.B. (1965) 'Kidney and marrow fat as indicators of white-tailed deer condition', *J. Wildl. Mgmt.* **29,** 397–8.
—— (1967) 'Reproductive biology of white-tailed deer in Manitoba', *J. Wildl. Mgmt* **31**, 114–23.
Ratcliffe, P.R. (1984a) 'Population density and reproduction of red deer in Scottish commercial forests', *Acta Zool. Fennica* **172**, 191–2.
—— (1984b) 'Population dynamics of red deer (*Cervus elaphus* L.) in Scottish commercial forests', *Proc. Roy. Soc. Edinburgh.* **82** B, 291–302.
—— (1987) 'Distribution and current status of sika deer, *Cervus nippon*, in Great Britain', *Mammal Review* **17**, 39–58.
Ratcliffe, P.R., and Rowe, J.J. (1985) 'A biological basis for managing red and roe deer in British forests, *Proc. XVII Congr.* IUGB, 917–25.
Ratcliffe, P.R, and Staines, B.W. (1982) 'Red deer in woodlands: research findings'. In: *Roe and Red Deer in British Forestry*, pp. 42–53. British Deer Society.
Reardon, P.O., Leinweber, C.L., and Merrill, L.B. (1972) 'The effects of bovine saliva on grasses', *J. Anim. Sci.* **34**, 897–8.
Riney, T. (1964) 'The impact of introductions of large herbivores on the tropical environment', *Int. Union Conserv. Nat. Public. New Ser.* **4**, 261–73.
Robinette, W.L., and Gashwiler, J.S. (1950) 'Breeding season, productivity and fawning period of the mule deer in Utah', *J. Wildl. Mgmt* **14**, 457–69.
Robinette, W.L., Gashwiler, J.S., Jones, D.A., and Crane, H.S. (1955) 'Fertility of mule deer in Utah', *J. Wildl. Mgmt* **19**, 115–36.
Robinette, W.L., Gashwiler, J.S., Low, J.B., and Jones, D.A. (1957) 'Differential mortality by sex and age among mule deer', *J. Wildl. Mgmt* **21**, 1–16.
Robinette, W.L., Hancock, N.V., and Jones, D.A. (1977) *The Oak Creek Mule Deer Herd in Utah.* Resource Publication, 77–15, Utah Division of Wildlife, 1–148.
Rudge, A.B. (ed.) (1984) *The Capture and Handling of Deer.* Nature Conservancy Council, Peterborough.
Sadleir, R.M. (1979) 'Energy and protein intake in reaction to growth of suckling black-tailed deer fawns', *Can. J. Zool.* **58**, 1347–54.
Santiapillai, C., Chambers, M.R., and Jayawardene, C. (1981) 'Observations on the sambar, *Cervus unicolor* Kerr, 1792 (Mammalia: Cervidae) in the Ruhuna National Park, Sri Lanka', *Ceylon J. Sci.* (Bio. Sci.) **14**, 193–205.
Schaaf, D. (1978) 'Some aspects of the ecology of the swamp deer or barasingha (*Cervus d. duvauceli*) in Nepal', In: *Threatened Deer*, pp. 66–86. IUCN, Switzerland.
Schaal, A. (1982) Influence de l'environnement sur les composantes du groupe social chez le daim, *Cervus (Dama) dama* L. *La Terre et la Vie, Revue d'Ecologie,* **36**, 161–74.
—— (1987) Le polymorphisme du comportement reproducteur chez le daim d'Europe (*Dama d. dama*). PhD thesis, Université Louis Pasteur, Strasbourg.
Schaller, G. (1967) *The Deer and the Tiger.* Chicago University Press.
—— (1972) *The Serengeti Lion – A Study of Predator-prey Relations.* Chicago University Press.
Scheffer, V.B. (1951) 'The rise and fall of a reindeer herd', *Sci. Monthly* **73**, 356–62.
Schladweiler, P. (1976) *Effect of Coyote Predation on Big Game Populations in Montana.* Project report W–120–R.7. Montana Dept Fish and Game, 1–26.

Seth, S.D.S., Mukhopadhyay, A., Bagotu, N., Prabhaker, M.C., and Arora, R.B. (1973) 'Antihistamine and spasmolytic effect of musk', *Japan J. Pharmacol.* **23**, 673–9.
Short, H.L., Newsom, J.D., McCoy, G.L., and Fowler, J.F. (1969) 'Effects of nutrition and climate on southern deer', *Trans. 34th N. Amer. Wildlife Conference,* 137–45.
Silliman, R.P., and Gutsell, J.S. (1958) 'Experimental exploitation of fish populations', *Fish Bull.*, US, **58**, 214–52.
Silver, H., Colovos, N.F., Holter, J.B., and Hayes, H.H. (1969) 'Fasting metabolism of white-tailed deer', *J. Wildl. Mgmt* **33**, 490–8.
Simpson, G.G. (1945) 'The principles of classification and a classification of mammals', *Bull. Am. Mus. Nat. Hist.* **85**, 1–114.
Sinclair, A.R.E. (1977) *The African Buffalo – a Study of Resource Limitation of Populations.* Chicago University Press.
—— (1979) 'Dynamics of the Serengeti ecosystem'. In: *Serengeti: Dynamics of an Ecosystem,* eds. Sinclair, E.R.E., and Norton-Griffiths, M., pp. 1–30 Chicago University Press.
Sinclair, A.R.E., and Norton-Griffiths, M. (1979) *Serengeti: Dynamics of an Ecosystem.* Chicago University Press.
Singh, A. (1978) 'The status of the swamp deer (*Cervus d. duvauceli*) in the Dudhwa National Park'. In: *Threatened Deer*, pp. 132–. IUCN, Switzerland.
Slobodkin, L.B. (1968) 'How to be a predator', *Amer. Zool,* **8**, 43–57.
Smith, R.H. (1979) 'On selection for inbreeding in polygynous animals', *Heredity* **43**, 205–11.
Spedding, C.R.W. (1971) *Grassland Ecology.* Clarendon Press, Oxford.
Staines, B.W. (1970) The management and dispersion of a red deer population in Glen Dye, Kincardineshire. PhD thesis, University of Aberdeen.
—— (1974) 'A review of factors affecting deer dispersion and their relevance to management', *Mammal Review* **4**, 79–91.
—— (1978) 'The dynamics and performance of a declining population of red deer (*Cervus elaphus*)', *J. Zool. Lond.* **184**, 403–19.
—— (1985) 'Humane control of deer in rural areas'. In: *Humane Control of Land Mammals and Birds*, pp. 105–10. Universities Federation for Animal Welfare.
Staines, B.W., Crisp, J.M., and Parish, T. (1982) 'Differences in the quality of food eaten by red deer (*Cervus elaphus*) stags and hinds in winter', *J. Appl. Ecol* **19**, 65–77.
Staines, B.W., and Welch, D. (1984) 'Habitat selection and impact of red deer (*Cervus elaphus*) and roe deer (*Capreolus capreolus*) in a Sitka spruce plantation', *Proc. Roy. Soc. Edinburgh* **82**B, 303–19.
Sterba, O., and Klusak, K. (1984) 'Reproductive biology of fallow deer, *Dama dama', Acta Scientarum Naturalium Academiae Scientarum Bohomoslovacae Brn XVIII* **6**, 1–52.
Strandgaard, H. (1972) 'The roe deer population at Kalo and the factors regulating its size', *Danish Rev. Game Biol.* **7**, 1–205.
Stubblefield, S.S., Warren, R.J., and Murphy, B.R. (1986) 'Hybridisation of free-ranging white-tailed and mule deer in Texas', *J. Wildl. Mgmt* **50**, 688–90.
Stumpf, W.A., and Mohr, C.O. (1982) 'Linearity of home ranges of California mice and other mammals', *J. Wildl. Mgmt* **26**, 149–54.
Suttie, J.M. (1979) 'The effect of antler removal on dominance and fighting behaviour in farmed red deer stags', *J. Zool. Lond.* **190**, 217–24.
Taber, R.D., and Dasmann, R.F. (1954) 'A sex difference in mortality in young Columbian black-tailed deer', *J. Wildl. Mgmt* **18**, 309–15.
Takatsuki, S. (1983) 'Group size of sika deer in relation to habitat type on Kinkazan Island', *Jap. J. Ecol.* **33**, 419–25.
—— (1987) 'The general status of sika deer in Japan', *Deer* **7**, 70–2.
Taneja, V., Siddiqui, H.H., and Arora, R.B. (1973) 'Studies on the anti-inflammatory activity of *Moschus moschiferus* musk and its possible mode of action', *Indian. J. Physiol. Pharmacol.* **17**, 241–7.
Tansley, A.R. (1922) 'Studies of the vegetation of the English chalk II. Early stages of the redevelopment of woody vegetation in chalk grassland', *J. Ecol.* **10**, 168–77.
Tansley, A.R., and Adamson, R.W. (1925) 'Studies of the vegetation of the English chalk III. The chalk grasslands of the Hampshire – Sussex border', *J. Ecol.* **13**, 177–223.
Taylor, W.P. (ed.) (1956) *The Deer of North America.* Stackpole Books, Pennsylvania.
Teer, J.G. (1984) 'White-tailed deer population management: lessons from the Llano Basin, Texas'. In: *White-tailed Deer: Ecology and Management,* ed. Halls, L.K., pp. 261–92. Stackpole Books, Pennsylvania.
Thalen, D.C.P. (1984) 'Large mammals as tools in the conservation of diverse habitats', *Acta Zool. Fennica.* **172**, 159–63.

Thirgood, S.J. (in prep.) Variation in social and sexual strategies of fallow deer. PhD thesis in preparation, University of Southampton.
Thomas, J.W., and Toweill, D.E. (1982) *Elk of North America: Ecology and Management.* Wildlife Management Institute/Stackpole Books, Pennsylvania.
Thompson, d'Arcy (1942) *On Growth and Form.* Cambridge University Press .
Topinski, P. (1974) 'The role of antlers in establishment of the red deer herd hierarchy', *Acta Theriologica* **19**, 509–14.
Trainer, C.E. (1975) 'Direct causes of mortality in mule deer fawns during summer and winter periods on Steens Mountain, Oregon', *Proc. West. Assoc. Game and Fish Commissioners* **55**, 163–70.
Trivers, R.L. (1971) 'The evolution of reciprocal altruism', *Qu. Rev. Biol.* **46**, 35–57.
Tudge, C. (1987) 'Custom-built deer take to the hills', *New Scientist*, 9th April, 1987, p. 28.
Van Ballenberghe, V. (1985) 'Wolf predation on caribou: the Nelchina herd case history', *J. Wildl. Mgmt* **49**, 711–20.
Van de Veen, H.E. (1979) Food selection and habitat use in the red deer (*Cervus elaphus* L.). PhD thesis, Rijksuniversiteit te Gröningen.
Verme, L.J. (1962) 'Mortality of white-tailed deer fawns in relation to nutrition', *Proc. Nat. White-tailed Deer Disease Symp.* **1**, 15–38.
—— (1965) 'Reproduction studies in penned white-tailed deer', *J. Wildl. Mgmt* **29**, 74–9.
—— (1967) 'Influence of experimental diets on white-tailed deer reproduction', *Trans. N. Am. Wildl. and Natur. Resources Conf.* **32**, 405–20.
—— (1969) 'Reproductive patterns of white-tailed deer related to nutritional plane', *J. Wildl. Mgmt* **33**, 881–7.
—— (1983) 'Sex-ratio variation in *Odocoileus:* a critical review', *J. Wildl. Mgmt* **47**, 573–82.
Verner, J. (1977) 'On the adaptive significance of territoriality', *Amer. Nat.* **111**, 769–75.
Vickery, P.J. (1972) 'Grazing and net primary production of a temperate grassland', *J. Appl. Ecol.* **9**, 307–14.
Vittoria, A., and Rendina, N. (1960) 'Fattori condizionanti la funzionalita tiaminica in piante superiori e cenni sugli effetti dell bocca del ruminanti sull erbe pascolative', *Acta Medica Veterinaria (Naples)* **6**, 379–405.
Walmo, O.C. (ed.) (1981) *Mule and Black-tailed Deer of North America.* Wildlife Institute of America, University of Nebraska Press.
Waterfield, M.R. (1986) Observations on the ecology and behaviour of fallow deer (*Dama dama* L.) PhD thesis, University of Exeter.
Watson, A., and Staines, B.W. (1978) 'Differences in the quality of wintering areas used by male and female red deer (*Cervus elaphus*) in Aberdeenshire', *J. Zool. Lond.* **286**, 544–50.
Watt, K.E.F. (1955) 'Studies on population productivity', *Ecological Monographs* **25**, 269–90.
Weiner, J. (1977) 'Energy metabolism of the roe deer', *Acta Theriologica* **22**, 3–34.
White, M.F., Knowlton, F.F., and Glazener, W.C. (1972) 'Effects of dam-newborn fawn behaviour on capture and mortality', *J. Wildl. Mgmt* **36**, 897–906.
Whitehead, G.K. (1972) *Deer of the World.* Constable, London.
Wiegert, R.G., and Evans, F.C. (1967) 'Investigations of secondary productivity in grasslands'. In: *Secondary Productivity in Terrestrial Ecosystems*; Vol. 2, ed. Petrusewicz. K., pp. 499–518. Warsaw.
Wislocki, G.B., Aub, J.C., and Waldo, C.M. (1947) 'The effects of gonadectomy and the administration of testosterone propionate on the growth of antlers in the male and female deer', *Endocrinol.* **40**, 220–4.
Wolff, J.O. (1978) 'Burning and browsing effects on willow growth in interior Alaska', *J. Wildl. Mgmt* **42**, 135–40.
Wood, A.J., Cowan, I.McT., and Nordan, H.C. (1962) 'Periodicity of growth in ungulates as shown by deer of the genus *Odocoileus*', *Can. J. Zool.* **40**, 593–603.
Woolf, A., and Harder, J.D. (1979) 'Population dynamics of a captive white-tailed deer herd with emphasis on reproduction and mortality', *Wildlife Monographs* **67**, 1–53.
Yanushko, P.A. (1957) 'The way of life of the Crimean deer and their influence on the natural cycle', *Trans. Moscow Soc. Nat.* **35**, 39–52.
Zejda, J., and Homolka, M. (1980) 'Habitat selection and population density of field roe deer (*Capreolus capreolus*) outside the growing season', *Folia Zoologica* (Brno) **29**, 107–15.
Zuckerman, S. (1953) 'The breeding season of mammals in captivity', *Proc. Zool. Soc. Lond.* **122**, 827–950.

Index